AF497483

To Shawkat and Micheline

ISBN: 978-2-9591693-4-2

Paperback

Edition N° 1

Editing: *Ranya Younes*

Chapter covers: *Darine Youssef*

Book cover & design: *Marilynn Bou Habib*

Table of Contents

Preamble

This book is about space. It is the story of the Universe told in my own way. By the end of this book, you will be left with the knowledge needed to understand what we know about space. So why should you read it? After all, everything happening in outer space might not affect your life today. Unless you are a professional astronomer, the knowledge of space will not inflate your bank account, help you survive, or improve your love life. Well, this last point is debatable; nothing can beat a romantic evening under the stars. Think of History, Art, and everything we label "culture". What do you get from visiting a museum or an art gallery? If you are not going to buy, sell, or deal with art in general that is. Well, apply the same concept to this book. Those who deal with space sciences have been building the museum of space knowledge, and this book is dedicated to giving you a quick, thorough tour of that museum.

History, geography, medicine, law, politics, finance, economy, arts, management, engineering, and all other sciences are based on knowledge and skills that we have acquired on one planet. One planet out of eight planets, revolving around one star out of 400 billion stars, in a galaxy of trillions of galaxies, in one Universe that might be one of who knows how many others. If these numbers are correct, Earth would be as big as a fish in the ocean. All the sciences today come from this little fish in that vast ocean. These sciences are essential and they occupy 100% of our time. Wouldn't it be worth spending a few hours out of a lifetime to discover the vast ocean that surrounds us? This book is for those of you who are curious to know more. For those seeking knowledge and entertainment. So, sit back and enjoy this captivating and insightful journey we will embark on together.

People always like to compare the mystery of space to that of the underwater. The ocean is vast, and the secrets of the underwater are undoubtedly the most thrilling. Let's stand on the water surface above the deepest ocean point, the Mariana Trench, and look down. There are 11 kilometers (km) of water under your feet, and a largely unexplored world. 11 km is 13 stacked Burj Khalifa towers, or 33 stacked Eiffel towers. We have tools like radars, submarines, and other vehicles to explore and visit the bottom of this ocean. However, it remains too large to examine entirely, and we still discover something new every day. Let's look up into the night sky now. Do you see this bright object in the sky reflecting on the water? That's the Moon. It is 384,400 km away, equivalent to around 460,000 (rounding down) stacked Burj Khalifa towers, or more than 1 million Eiffel towers. Do you see the second brightest object in the night sky that looks like a star? It is not a star; it is Venus. It is one of our closest neighboring planets, and it is currently 280 times farther than the moon (but this number can change). The Sun is around 390 times farther, but it is also 1 million times larger than Earth. The brightest star that we see is astonishingly 200 million times farther. When it is dark, you can see a dim cloud-like shape in the sky, and this is our closest galactic neighbor, Andromeda. It is 60 trillion (12 zeros) times farther than the Moon. It is a galaxy that hosts 1 trillion stars, more or less like our Sun. This is literally our neighborhood. Do you see where I am going with this? Do you see the things we will explore together?

What is new in this book? For thousands of years, Astronomers used optical light, that is light that we humans can see, to study the Universe. Nowadays, we study the Universe through light across the whole electromagnetic spectrum. Those range from radio waves, microwaves, infrared radiation, optical light, ultraviolet radiation, X-rays, and gamma rays. We also investigate other particles we receive from space, like neutrinos and cosmic rays. Very recently, we also started using gravitational waves. All these are called messengers because they carry information from space. Today, we study the Universe through multi-messengers. We look up, and sometimes look underground, to conduct experiments that help us identify the different particles in our Universe. We are at an advanced stage of space exploration, yet we are still far from interstellar and intergalactic travels. I wrote this book to combine the knowledge of space that we have gathered until today, 2021-2023, and recorded it in the simplest way possible. Be patient! With every new chapter, your view of the Universe will change, and your understanding of space will slowly, but surely, grow.

This book will also give you an insight into how discoveries are made and how scientists study space.

Before we start, I would like to refer to an interview with a famous scientist, Richard Feynman. When asked why two magnets repel when brought together, Feynman spent a few minutes explaining that answering a "why" question is not easy since the question can run very deep. He used this example: Aunt Many is in the hospital. Why? She slipped on ice and broke her hip. This could be a satisfying answer for most people. However, it may not satisfy someone from another planet who is unfamiliar with earthly concepts. They might ask: Why is ice slippery while other solids are not? Why is aunt Many taken to a hospital and not somewhere else? How did she get to the hospital? Why did this man drive her there? Why does he care about her? Are we going to explain marriage, love, and affection to answer why aunt Many is in the hospital? We could. How is affection created? By chemicals in our brains? Why does our brain produce these chemicals? How are the substances bound together? So, how do we answer the 'why' questions?

This book aims to satisfy the reader's curiosity. It is for everybody who would like to learn about space but may not be a professional in the field. Professionals are also welcome as the book provides a general picture of space. To provide satisfactory answers compatible with all readers, for most topics, I will keep the details brief, and at a 101-level course.

The Basic Questions: Why, What, Who, How, When, and Where

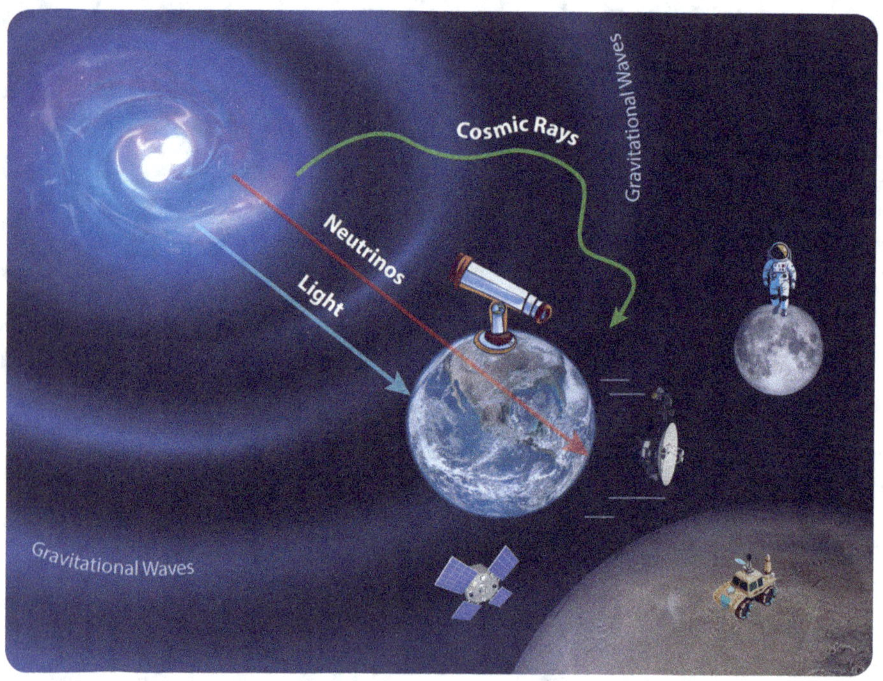

Cosmic Rays
Gravitational Waves
Neutrinos
Light
Gravitational Waves

Astronomy is the study of celestial objects, particularly their positions and movements. The word Astronomy is used for the study of Space and the Universe in general. It is a large scientific field, under which specific areas of space sciences lie, like Astrophysics. Astrophysics is the study of physical phenomena in the Universe. In particular, the physical phenomena behind the positioning, movement, creation, evolution, and death of celestial objects. Astrology, on the other hand, studies the movement and positioning of celestial objects with an interpretation of their influence on human behavior, character, and day-to-day life. I am not a believer in Astrological sciences, and like the majority of scientists, I do not believe that it has solid evidence to be considered a reliable tool to study space. This brief mention of Astrology in my book is, in my opinion, more than enough given its merits. This book will focus on Astronomy and Astrophysics in particular.

Why do we Study Space?

I believe there are five reasons to study space. The first is the curiosity to explore and to understand the grand design. Curiosity is innate, and it is what drives us forward. Finding answers to questions is a motive for humankind. Everywhere we look, there are mysteries and questions awaiting answers. For example, why is the sky blue? Why do stars twinkle at night? I will actually answer those questions later on.

The second reason to study space is for the general understanding of life and physics. Many of the experiments required to understand certain physics concepts cannot be done on Earth because we are limited by space and resources most of the time. Therefore, we look at the skies to study different objects that can, for instance, energize matter to levels we cannot reach on Earth and at scales far beyond the solar system. Once we understand these objects and the physical concepts behind their behavior, we understand physics. When we understand physics and how things work around us, we unlock new technologies and open the doors to the prosperity and advancement of the human race. We have heard of the relationship between Einstein's general relativity and the Global Positioning System (GPS). In a nutshell, if we did not know general relativity and account for its implications to satellite orbits and triangulation, we would have, at best, a localization precision of a few kilometers. I don't want my GPS to tell me that I am in Paris — thank you, but I already know that. I need it to tell me precisely in which part of which street I am now, the direction I should head to, and when I will arrive at the bar where my friends are waiting. Moreover, some objects in the Universe are more or less advanced in their stage of evolution than our planet and our system. By studying these objects, we can indirectly explore what our past and possible future look like.

The third reason to study space is convenience; the tools we develop for space exploration can be at the service of humankind. Computers, cameras, communication devices, medical imaging devices, and many more devices owe a part of their development to space exploration. Many of the tools we use daily were once developed or optimized for space exploration. For example, cordless power tools were first invented for Astronauts in servicing missions. As it is difficult to find a 220 Volt (V) socket in space, cordless tools were created. Nowadays, cordless tools are used everywhere and are present in every household. More directly, everything related to satellites, such as communication, navigation, weather and climate studies, would not exist without the satellites in space. Of course, space exploration is not the only innovative science, but it is definitely in the top list of sciences that are developing leading-edge technologies. With a dedicated budget, that very rarely exceeds 0.1% of the total income of a country, the innovations and advancements of space exploration are impressive.

The fourth reason to study space is cultural. Space is art, and like all the other fields, it has beauty. Space is also nature, as it examines the wonders of the Universe. Hundreds of space museums and planetariums built, and hundreds of thousands of annual visits to space testify to its need.

The fifth, and final reason to study space is that it is a must. There will come a point when life will no longer be bound to one single planet. Do you think that we would want to stay on the same planet forever? In the long run, we might have dozens of reasons to expand to other worlds. They can range from seeking pleasure, exploration, curiosity, and possibly the need to search for new resources, and, evidently, to avoid extinction. We might inevitably need to leave our home one day (if we manage to survive self-extinction threats on Earth in the upcoming years). The physics and technology reachable today might not be enough to sustain us, and we need to understand a lot of what is going on. So we have been looking for answers in space.

What to Study and Who Studies Space?

We are living on a planet, Earth, that is rotating around a star, the Sun. Seven additional planets orbit the same star alongside other objects like asteroids, comets, and other small planetary objects. Most of these planets have moons orbiting them. The Sun is one of around 400 billion other stars in what we call the Milky Way; our galaxy. It is estimated that more than half the stars in our galaxy have planets that orbit them. You can think of a galaxy as a collection of stars and other things, such as gas and dust, that are bound together by gravity. There are many trillion galaxies forming what we know as the Universe. The Universe also contains other things, like dark matter and dark energy, that we do not fully understand yet. In the upcoming chapters, I will try to explain every concept mentioned in this paragraph.

Space sciences are vast and diverse. Like medicine and engineering, one would have to pick an area of expertise. To cite a few areas:

- **Planetary Astrophysics** deals with planets and other objects that orbit the stars. It focuses on their characteristics, compositions, and evolution; with a particular interest in the solar system, and Earth.
- **Stellar Astrophysics** deals with stars. It focuses on their composition, movement, and mainly their evolution from the time they are born until their death; with a particular interest in our star, the Sun.
- **Interstellar Astrophysics** deals with what is found in between stellar (star) systems, which looks like a void but is not.
- **Galactic Astrophysics** deals with galaxies. It focuses on their evolution, interactions, and their clustering and distribution in the Universe; with a particular interest in our galaxy, the Milky Way.
- **Transient (or Time-domain) Astrophysics** deals with the changing phenomena in the Universe such as the emergence of a comet in our solar system, the explosion of stars, and the merger of black holes... but what is a black hole...?

- **High-energy Astrophysics** deals with everything that can generate high-energy particles. Mainly cataclysmic phenomena, like star explosions, black holes, galaxy core eruptions, etc.
- **Astroparticle Physics** (sometimes referred to as Particle Astrophysics) is a close cousin to high-energy astrophysics and deals with the elementary particles in astrophysical objects and their relationship with astrophysical phenomena.
- **Cosmology** deals with the shape, the beginning, the composition the evolution and the end of the Universe as a whole.
- **Astrobiology** deals with life in space and extraterrestrial forms of life.

These are a few of the main areas of Astrophysics. However, other areas of expertise also deal directly and indirectly with space. I am not talking about math (calculus, algebra, geometry, trigonometry, statistics, probabilities, etc.), which, in the end, is also a tool for studying physics and science in general. Theoretical physics lays the ground for the theories we use and investigate to understand the Universe.

Nowadays, physics is divided into two main branches. *Quantum physics* deals with everything on the microscopic level, and *general relativity* deals with the macroscopic aspects.

Four forces in the world exist: the electromagnetic force, the weak force, the strong force, and the gravitational force. At this point, one might ask, why only four? Pushing my friend is a force. My hand is exerting a force on my friend's shoulder. Why is pushing not in the four forces of the Universe? The answer is that while your hand exercises a force on your friend's shoulder, it is not fundamental. It is, however, the product of the fundamental forces that occur on the microscopic level between the atoms in your hand and your friend's shoulder. These fundamental forces are the ones we measure; they are in the list of the four fundamental forces. Scientists today are working on a theory of everything that unifies quantum physics and general relativity that can merge the four known forces. The first three forces are unified, however, gravity is a difficult fit in the equation. You will find in later chapters that gravity plays a significant role in the determination of the evolution of objects in the Universe, and the evolution of the Universe in itself.

Engineers play an important role in building the tools needed to visit and explore space. Aerospace engineers create the engines and vehicles, such as rockets and probes, that are essential for space missions. Both engineers and scientists work together on designing every component and programming each phase, from take-off to landing.

Other disciplines also play important roles in space exploration. Instrumentation develops the needed scientific tools, like telescopes, detectors, and cameras, used to collect valuable data and make the needed discoveries. Computer science uses data for analysis and simulations. The field of tailoring is used to design astronaut suits. Movie productions portray space adventures through sci-fi films.

Transportation experts move different parts of a space mission to its assembly areas when needed. Security and cybersecurity protect missions from external threats. Architectural experts plan for extraterrestrial colonies. Law professionals establish rules and legislations governing space conquest. Biology, medicine, and psychology help address physiological and psychological challenges in space.

How do we Study Space?

To study an object in space, like a planet or a star, we can either go there, send robots and probes, or look at the messengers we receive from this object. Going there is not a very efficient method nowadays. The only place we've been to, other than Earth, is the Moon, and the last mission to the Moon was in the 1970s. We have sent probes to all the planets in our solar system and sometimes landed rovers on their surfaces. Sending a rover or a probe with current technology is only achievable for planets in our solar system or nearby objects. This leaves studying the messengers that we receive from space objects.

A messenger is an entity, a particle, or an emission, that carries information. For example, information that optical light can carry is the object's color. The blue light emerging from my friend's T-shirt into my eyes indicates the color of the T-shirt: blue. The light in this scenario is a messenger that brings me information; in this case, the color.

Each class or subclass of messengers is created by specific physical processes. Sometimes we need to combine information from several messengers to learn what is happening inside a particular astronomical source. We can deduce the physical processes at the origin of these emissions by analyzing the type, amount, and composition of each messenger received on Earth. Contrary to general belief, to study distant objects, astronomers do not often have access to detailed surface images of the object, and have even less access to under-the-surface images. However, astronomers can do wonders only by looking at the type and the amount of messengers received.

From an insider's perspective, it goes like this: theoreticians study a particular source and create simulations and models that predict the type of messenger, and the amount that should be detected from this kind of source. These models, based on theories, are nowadays mostly generated with the use of computers, and often these predictions can change with time. Several models can describe a particular phenomenon, and sometimes there is no way to determine which one is right. Instrumentalists and engineers build complex and advanced instruments capable of detecting messengers from the sky. Experimentalists and observers then use these instruments to collect and analyze the data using computers. The data is then compared to the models and the theoreticians' predictions to see which model fits, and which is discarded. After this, we can know precisely what is happening inside and around a source in the sky. This is usually how discoveries are made.

Suppose we are talking about the explosion of a star. In this case, the light detected at the beginning of the explosion is expected to brighten and then dim with time in a specific way. Experimentalists using telescopes will often see a new source in the sky that will become brighter for a certain amount of time before it gradually fades away. They compare the behavior of the light they saw to the predictions of the theoreticians and find that the data matches exactly the model of an exploding star. They then conclude that what they saw was the explosion of a star, a supernova. They did not see the actual explosion; they did not see the star shatter in a dazzling display of fireworks. They just saw the amount of light from a source change with time.

In some cases, observational data can be unexpected and it can surprise scientists as it can lead to something new that was not accounted for. Other times, observational data can show something new but accounted for. So, the importance of a finding can vary. For example, there are millions of asteroids in the solar system, so discovering a new asteroid in the asteroid belt is undoubtedly new, but not surprising. Finding an asteroid on a collision path with Earth is new, but not surprising; it is however significantly important. Discovering that the asteroid's movement is not entirely governed by the laws of physics is new, surprising, and very important. In general, models, theories, and knowledge can be built around the information brought by new observations. These scientific methods extend from finding new objects to discovering new physical models and theories. We know of four types of messengers.

The first type of messenger is electromagnetic radiation. Optical light is electromagnetic radiation, but it is only a small fraction of the electromagnetic spectrum. Different types of electromagnetic radiation are characterized by their energy. The lowest energies can be found in the radio waves domain. After that, we have microwaves, infrared radiation, optical light, ultraviolet radiation, X-rays, and gamma rays.

Why do we need to study all that? Well, just like for sound, the human eye can only see a small portion of electromagnetic radiation. Our ears can hear sounds between 20 Hertz (Hz) and 20 kilohertz (KHz). Dogs, for instance, can hear ultrasounds up to 60 KHz. These are more acute sounds that dogs are more responsive to, such as acute whistling. Some scientists also suggest that dogs can hear earthquakes before we do because they hear the sound produced by the ground, below 20 Hz, that humans cannot detect. This means that the information is there, even though we cannot access it with our human senses. The same goes for the electromagnetic spectrum. The optical light, which is commonly seen by humans, is divided into six colors: red, orange, yellow, green, blue, and violet. Even though we can only see optical light, other information is contained in different electromagnetic ranges. The data is there, but we don't have direct access to it. This is why we spend so many resources such as money and time to develop and build instruments that detect these different ranges. This is not only applied in astronomy. The military, for example, often uses special goggles for detecting infrared radiation to see in the dark. Infrared radiation is emitted by the heat of the human body, even in darkness. Therefore, soldiers equipped with infrared goggles can see this radiation and see people in the dark.

One or more physical processes produce one or more electromagnetic radiation types. Detecting the resulting radiation gives us access to the physical process behind it. Nowadays, we detect optical light from space using telescopes. These telescopes can be either on the planet's surface or in space. The idea is always the same: a mirror, or a lens, reflects, or refracts, light from space into a camera that captures and records this light. The bigger the mirror or the lens, the more it can accumulate light. The more light is collected the more information is revealed. In this scenario, we can imagine light particles as rain droplets and the mirror as a bucket. The larger the bucket, the more we can accumulate water. With more water, we have more data and thus better access to information for better results.

We detect radio waves and microwaves using antennas or big metal dishes. We use metal dishes because radio waves can be reflected by metal. We detect infrared and ultraviolet light with special cameras mounted on special telescopes, similar to optical telescopes. The scopes of the camera and the telescope mirrors are what differ here. X-rays and gamma rays are harmful to the human body, and luckily, our atmosphere blocks them. Therefore, we have to send special detectors to outer space to be able to see these radiations. Very high-energy gamma rays can be detected on Earth, but I will not bother you with more details on that now. Instead, I will later dedicate a chapter to describe the different instruments used in astronomy and space sciences.

The second type of messenger is cosmic rays. These are charged particles emitted by space objects that enter and interact with our atmosphere. The primary source of cosmic rays on Earth is the Sun. The Sun mostly emits electrons, positrons, protons and helium nuclei that travel toward Earth. The planet is shielded by a magnetic field that deflects these charged particles toward the two poles. When these particles collide with other particles in Earth's upper atmosphere, they create the magnificent phenomena known as northern and southern lights: Aurora Borealis and Aurora Australis, respectively. Other, more energetic, cosmic rays come from far beyond; either from different regions in the galaxy or from extragalactic origin. As mentioned, these particles are charged and deflected by the galaxy's magnetic field. Therefore, it is complicated to determine their initial source and direction. As their paths are diverted, cosmic rays don't travel in straight lines like the rest of the messengers; instead, they appear to come from places that are not their origin. Detecting these cosmic rays shows that we have particles, like iron cores, that can reach ultra-high energies that can only be produced in highly energetic cosmic events. While searching for cosmic rays, Carl Anderson discovered a positively charged electron. Since electrons were negatively charged, his discovery led to the detection of the first anti-matter particle, a positron. The positron is an anti-matter particle or an antiparticle with properties opposite to "regular" particle properties. This is the positron of the Positron Emission Tomography scan, known as PET scan which is used in medicine to detect and monitor cancers.

The third type of messenger is neutrinos. In the particle world, we can think of neutrinos as ghost particles. They are particles that hardly interact with anything else in the world. In other words, they can easily travel through any object in the Universe. To efficiently stop neutrinos, we will need to build a wall of lead (Pb) that is several trillion kilometers thick, which is highly impractical as the wall would be much thicker than the solar system. Billions of neutrinos travel through our bodies every second without any interaction. Billions are traveling through you as you are reading this sentence now. So how do we detect them? We would have to think in probability matter. There is always a minute chance that a neutrino will interact with another body on Earth; like a vast body of water or ice. Neutrino hunters build their detectors inside massive bodies of water; deep in the sea or the ice. The aim is to detect neutrinos' interaction with other particles and record their imprint, but even with all these efforts, only a handful of neutrinos are recorded. Detecting neutrinos allows scientists to investigate the specific physics laws implicated in their creation process. Although detecting neutrinos is very difficult, they can provide important information since they can escape places light cannot due to their shallow interactions. Such sites are, for example, the hearts of cataclysmic explosions of stars.

The fourth, and final, type of messenger is gravitational waves. To understand gravitational waves, we must understand Einstein's general theory of relativity. I will use this opportunity to introduce general relativity. I will not go into the details at this point, but I will use a typical example to explain the main idea. Relativity stipulates that time and space are interconnected in one fabric called space-time. It is a 4-dimensional (4D) fabric, made of 3 dimensions of space (height, width, and thickness), and 1 dimension of time. Space-time is the fabric of the Universe. We would need to imagine a 4D object to understand the concept of space-time, but since we cannot do that, let us remove one of the dimensions and think in a 3-dimensional (3D) manner. We have a 3D fabric that represents the fabric of the Universe. Everything we know; every object in the Universe, galaxy, black hole, star, planet, moon, asteroid, or comet, is submerged in this fabric. We can imagine this fabric as a water mattress. Imagine the Universe being a water mattress, and an object, like planet Earth, is a bowling ball on the surface of this mattress. Sitting on the mattress, the bowling ball will form a dent in the surface of the mattress as illustrated in Figure 1.1. The heavier the bowling ball is, the deeper the dent is. What happens if I take another lighter object, like a marble representing the Moon, and place it next to the bowling ball? The marble will fall in the dent toward the bowling ball. Einstein explains that this is how gravity works. Gravity is not a force; it is an illusion of a force of attraction created by heavy objects, like the Earth, that bend the space-time fabric allowing other objects to fall toward them. It has the effects of a force and is fundamental in itself, which is why it is considered the fourth force. Now if there is no friction on the mattress, and I kick the marble slightly away from Earth, the marble will move in circles around the bowling ball. This is why the Moon is in orbit around Earth. The same train of thought applies to objects like the Sun and the Earth, Jupiter and another moon, black holes, and every other object in the Universe.

Figure 1.1

Now let's talk about black holes. This time the bowling ball represents a black hole. If I use two bowling balls, I will have two big dents in the water mattress. What happens to the water if I rotate these two bowling balls on the mattress as if the black holes were spiraling toward each other? It creates waves. The waves travel in the water to the edges of the mattress, no matter how big it is. These are gravitational waves; distortions in the space-time fabric that travel through the Universe. Since Earth is also on the mattress, some of these waves reach us. Practically, the black holes must have different masses to efficiently generate gravitational waves. We have the technology today to detect these kinds of distortions, however, they must come from heavy objects like merging black holes, dead star cores, or very energetic events (more information on that later). When we detect this kind of distortion, it tells us that somewhere in the Universe, two black holes are rotating toward each other and merging into a bigger black hole. It also informs us of their specifics; like the mass of these objects, and the system's inclination. Gravitational waves are important as they bring information from places light can't escape, like the very beginnings of the Universe. Keep in mind that gravitational wave astronomy is only a few years old, and the domain has yet to reveal its major discoveries in astrophysics, cosmology, and physics.

All the information in the following is the result of studying these four messengers and combining the information we get from the skies. When different messengers are combined to study an object or a phenomenon, it is called multi-messenger astrophysics.

When and Where do we Study Space?

Every time we look at something, we see it in the past. This is because light needs time to travel from point A to point B. When I look at a distant street lamp, I receive the light emitted by the lamp. This light has to travel the distance between the lamp and me, which takes some time. I am looking at the lamp as it was when it emitted these light particles. Light takes a fraction of a second to reach my eyes. It is the speed of light multiplied by the distance. In this case, the delay is not noticeable because light is traveling a relatively short distance. On Earth, we deal with distances up to the kilometer scale. Light's speed is approximately (~) 300,000 kilometers per second (km/s) in the void, and the Earth's circumference is around 40,000 km. So, everywhere I look on Earth, I receive light in a fraction of a second, and the delay will not be noticeable. A beam of light can go around the Earth around 7 times in 1 second. However, this effect is much more pronounced in space.

So, let's talk about the concept of distance in space. The moon is 380,000 km away, and light from the moon takes a little over one second to reach my eyes. Therefore, I am looking at the moon as it was a second ago. Light from the Sun takes 8 minutes and 12 seconds to reach Earth. If the Sun were to disappear now, we would still have 8 minutes and 12 seconds of light on Earth before darkness falls. The distance between the Sun and Earth is 150 million km. In astronomy, we use this distance as a reference, calling it an astronomical unit. 1 astronomical unit (au) is roughly 150 million km. The distance between the orbits of Earth and Mars is 0.53 au; between Earth and Jupiter 4.2 au; between Earth and the farthest planet, Neptune, 29 au; and between Earth and the outer edge of the solar system 100,000 au. These are average distances. Remember that the planets constantly move, and the distances between them vary a lot.

The closest star to the Sun is Proxima Centaury, and it is at an astonishing ~ 40,170,300,000,000 km. At this scale, the use of km, or even au, is not convenient anymore. So, astronomers introduced the concept of light-years (ly). Light year is a unit of distance, not time. It is the distance that light travels in a void during one year. In other words, there are 31,557,600 seconds in a year, and for each one of these seconds, light travels around 300,000 km. During one year, light travels approximately 9.5 trillion kilometers, so 1 ly is equivalent to approximately 9.5 trillion kilometers. Proxima Centaury is 4.25 ly away. This means that light takes 4.25 years to reach Earth, from Proxima Centaury. We look at Proxima Centaury now as it was 4.25 years ago. If Proxima Centaury miraculously disappears now, we will not know before 4.25 years. If we were on a ship traveling at the speed of light (relatively impossible), it would take us 4.25 years to reach Proxima Centaury. With current space travel technology, it would take us more than 50,000 years to reach Proxima Centauri, because we travel at speeds much slower than the speed of light. 50,000 years is a considerable time. The first pyramids were built around 5,000 years ago, so it would take 10 times this time to reach Proxima. In comparison, we need around 6 hours to reach the Moon with our current rockets. Our galaxy is 100,000 ly wide. All the stars we see at night are within a few thousand light-years away. The closest galaxy to us is Andromeda, at 2.5 million light years. NGC 4845 is a galaxy 65 million light-years away. If someone in this galaxy were to point a gigantic telescope at Earth now, they would see how Earth was 65 million years ago. They would probably be able to see the dinosaurs' extinction! The size of the observable Universe is approximately 93 billion light years. For the rest of this book, keep in mind that light is the fastest thing that can move through space; nothing can travel faster.

Technical notes: Astronomers use the more scientific concept of parsec (pc) for distances in space. 1 pc is 3.26156 ly, but for convenience, we will stick to light years in this book. To calculate the distance to close objects, like stars in our galaxy, astronomers use the concept of parallax. They use the relative position of the stars as Earth is orbiting the Sun. To understand the concept of parallax, place a finger in front of your face and close one eye. Open your eye and close the second one; your finger's position, relative to the background, changes. By applying basic trigonometry, you can calculate the distance to your finger.

As shown in Figure 1.2, positions and sizes in the sky can be measured by angles. The angular distance between two objects is an indication of how far they are from each other on the two-dimensional celestial sphere. The angle is composed of the two objects and the observer's position on Earth. Following this logic, the position of a nearby star will change in the sky as the Earth moves around the Sun. By comparing the two positions, we can calculate the distance of nearby stars. The angle formed by the two apparent positions of the star is called parallax. 1 pc is the distance to a star with a parallax angle of 1 arcsecond (An arcsecond is a unit of angles). 1 degree (°) is composed of 60 arcminutes, and 1 arcminute is composed of 60 arcseconds. The angular size of an object is an indication of how large the object appears on the 2D celestial sphere. It is composed of the two edges of the object with the observer on Earth. The angular size of the Moon, or the angular distance between two edges on the Moon circle in the sky, is half a degree, or 31 arcminutes to be precise. It is the Moon's angular diameter.

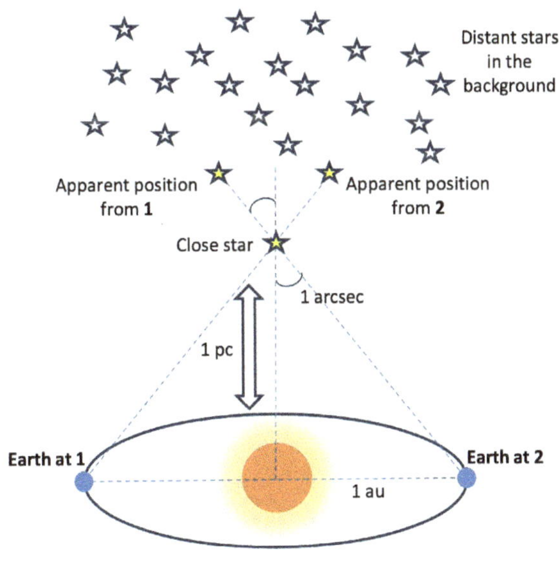

Figure 1.2

To calculate the distance of farther objects, we can use the properties of some stars. But before I explain this, you should know that one of the characteristics used in astronomy is brightness. The apparent brightness of a star is a measurement of how bright it appears regardless of its distance. It is just an indicator of the amount of light that reaches us. The absolute brightness of a star is its brightness, or what its brightness would be if it were at a distance of 10 pc. Absolute brightness is an absolute measure, while apparent brightness is a relative one. For example, a very bright star might have a low apparent brightness if it is very far, and vice versa. If we put this star at a distance of 10 pc, it would appear much brighter. In other words, absolute brightness is only dependent on the luminosity of a star, while apparent brightness is dependent on luminosity and distance. Back to distance measurement, Cepheid stars are variable stars (their light varies) and their variability is related to their luminosity or absolute brightness. By studying their variability, one can determine their absolute brightness, and by observing them, one can determine their apparent brightness.

By comparing their absolute brightness to their apparent brightness, one can determine their distance and, subsequently, the distance of the region where they exist. In fact, this is how Hubble determined that the Andromeda galaxy is outside the Milky Way, but more on that later.

Other concepts, such as redshift, are used for more distant objects. I will explain redshift later in this book. These techniques measure how the composition of the light (signal) received shifts relatively to determine how far the light source is.

History of Astronomy

This chapter describes the evolution of our perception of the Universe throughout history. It is a chapter that shows how we thought of the Universe from the days of Aristotle until today, 2021-2023. If you skip this chapter, you will still be able to understand the remaining ones. However, you will miss the fun of witnessing the evolution of ideas that have unraveled thousands of years ago into logical and scientific arguments and have led to our understanding of the grand design as it is today.

The Geocentric Model

A very long time ago, knowledge was limited to a handful of people, and the tools and means for exploration were basically non-existent compared to recent times. We can imagine that day-to-day life was mostly about providing food and shelter. I imagine it went something like this: wake up, take advantage of the daylight and warmth provided by the yellow ball of fire in the sky, get things done, and take shelter when the yellow ball goes down leaving behind coldness and darkness. Luckily, we have this other ball in the sky surrounded by small shiny dots that provide a bit of light during the night. However, for some, this was not enough. They started asking questions and seeking answers. What is controlling this yellow ball of fire in the sky? How is it moving? How does it work? Is it the same as the one we see at night? What are these tiny dots in the sky, and why do we only see them at night? What is the meaning of the different shapes they form?

It is believed that it was around 500 years Before Christ (BC) that Pythagoras first suggested the notion of a spherical Earth. The first widely used model of the Universe is the geocentric model, which places Earth in the center of the Universe with all the other bodies revolving around it. This was proposed by Eudoxus, a student of Plato, who was a student of Socrates. Aristotle, also a student of Plato, was a philosopher who created his model of the Universe in the 4th century BC centered around Earth's sphere. According to him, at the time, the Universe was finite in space but infinite in time. Aristotle's model explains that the objects in the sky, except comets, are outside Earth's sphere. They are in, what is called, the celestial sphere that surrounds Earth. Although there were other philosophers at the time who believed that Earth could not be the center of the Universe and that the Sun was a better fit, Aristotle's geocentric model remained dominant until a few centuries after Christ (AD), until the Middle Ages (15th century).

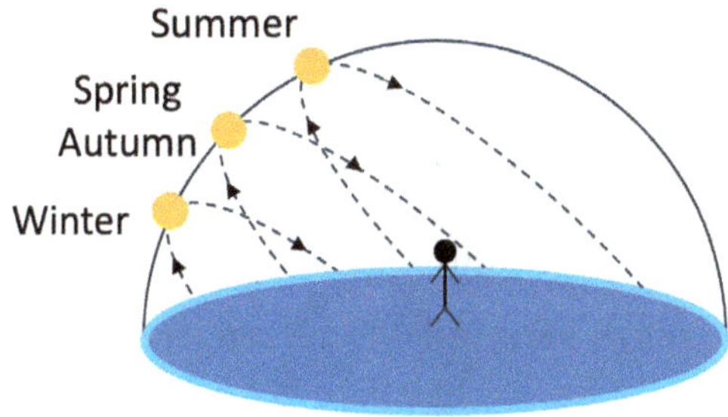

Figure 2.1

The geocentric model shows Earth as the center of the Universe, with a celestial sphere around it that contains stars and other objects, except comets. The position of the highest point of the Sun, at noon, in the celestial sphere changes throughout the year. It consists of the Ecliptic, which is a sinusoidal(S)-shaped path in the sky centered around the projection of the equator. This is in relation to the seasons. It is summer and warm when the Sun is at a high altitude, but when the Sun is at a low altitude in the sky, it is winter and cold. There are two solstice points at the beginning of summer and winter when the Sun reaches its maximum and minimum altitude, respectively, at noon. As shown in Figure 2.1, the summer solstice is when the Sun is the highest in the sky at noon, above the Tropic of Cancer line on the globe. It is the beginning of summer and the longest day in the Northern Hemisphere (around June 21st). However, at that time, it would be the beginning of winter and the shortest day in the Southern Hemisphere. The winter solstice is known as the day when the Sun is the highest in the sky at noon above the Tropic of Capricorn. It is the beginning of winter and the shortest day in the Northern Hemisphere (around December 21st), but it would be the beginning of summer and the longest day in the Southern Hemisphere. The start of Spring and Autumn are usually depicted by the equinoxes, where the Sun is directly overhead at the equator at noon, and the day and night are nearly equal in length (around March 21st and September 21st).

During the geocentric model times, many discoveries were made that later contributed to the rise of the heliocentric model (where the sun is the center of the Universe instead of Earth), and a better understanding of the grand design. From this, the estimation of the distance to the Sun and the Moon was made by Aristarchus in the 3rd century BC. Using trigonometry, and the fact that the Moon and the Sun have the same apparent size in the sky, Aristarchus estimated that the Sun is 20 times more distant than the Moon. This was the first rough estimation of the distance between the Earth and the Sun, and the Earth and the Moon. The Moon is much closer to Earth than the Sun. Today, we know that the Sun is approximately 386 times more distant from Earth than the Moon. Another noticeable discovery was Eratosthenes's estimation of the Earth's circumference around 240 BC. Eratosthenes measured the shadow that the Sun made on a sundial at the solstice in Alexandria; Egypt, and compared it to the shadow in another city, Syene, known today as Aswan. By knowing the distance between the two cities, and calculating the difference in the two measurements he assessed the circumference of Earth to be 40,320 km. The actual estimation today is 40,075 km. Not bad! Given the limited tools he had at the time. An example is given in Figure 2.2.

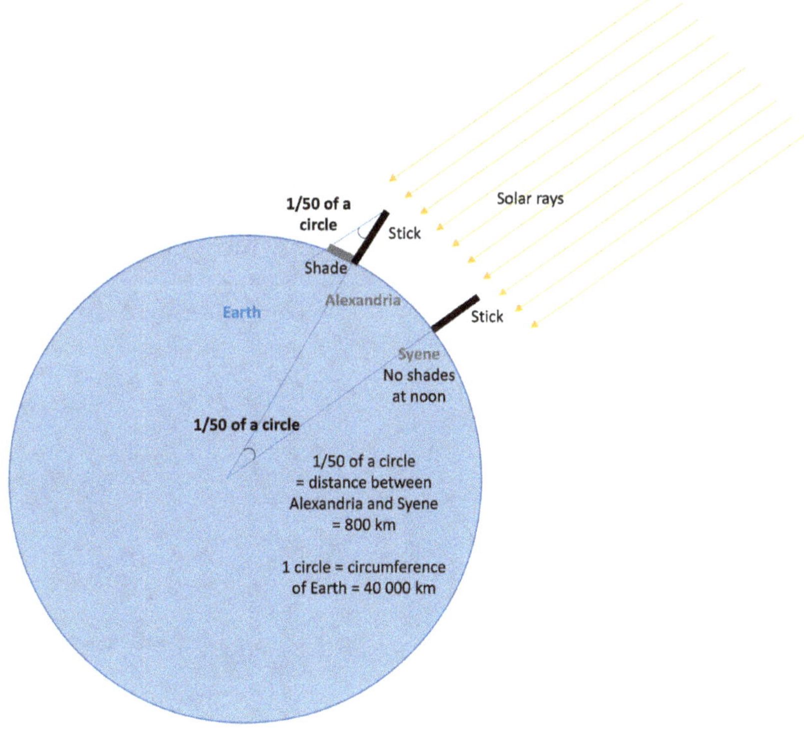

Figure 2.2

One thing to remember is that planets and stars look very similar in the night sky; they all appear as shiny dots. Planets were not known as planets, and the difference between a planet and a star was yet to be identified. One of the main phenomena that influenced the geocentric model conception is the retrograde — moving backward — motion of the planets (the real reason behind this phenomenon is shown in Figure 2.3). Venus is the second brightest object in the night sky, after the Moon, and is easy to find. It is usually found rising from the horizon, either at dawn or at dusk. If you observe Venus for several weeks, you will see that the planet's position drifts eastwards with time. It will then stop and drift in the opposite direction. This is the retrograde motion of Venus. Planets' retrograde motions are an optical phenomenon often called apparent retrograde motion. It is specific to planets and is caused by the relative position of the planets and their motion around the Sun. Venus retrograde is caused by the Earth swinging past the orbit of Venus. This apparent motion was very intriguing to geocentric model astronomers.

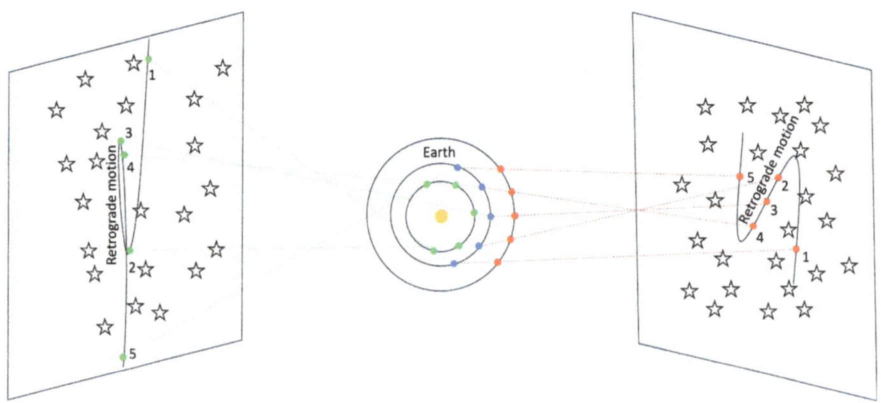

Figure 2.3

Planets were called wanderer stars because of their different behavior/motion. Ptolemy, a Greek astronomer born in 100 AD, tried to explain this apparent retrograde motion. He introduced the notion of epicycles to the geocentric models to explain the apparent retrograde motion. Epicycles are extra circular-shaped motions to the stars, in addition to their original orbits around Earth, which explains their apparent retrograde motion in the celestial sphere. The model, shown in Figure 2.4, was intricate but could account for the complex motion of the wanderer stars in the celestial sphere. However, the model had to reset every few hundred years. This model remained the leading model until the Middle Ages.

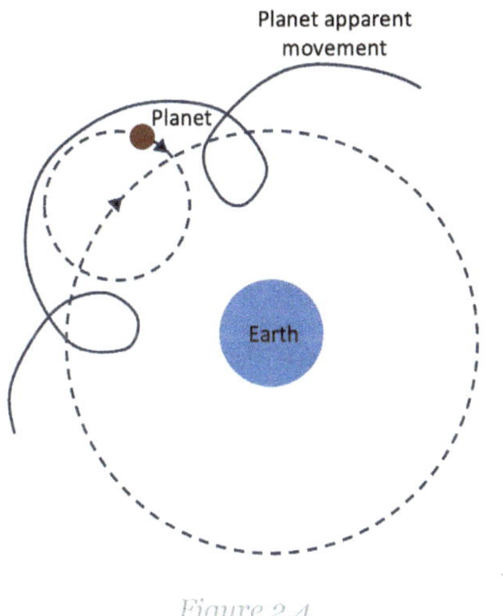

Figure 2.4

It is also worth mentioning that Ptolemy was the first Greek astronomer to catalog constellations. Constellations are groups of stars that appear to draw a particular shape in the sky. Of course, the apparent position of the stars in a 2D celestial sphere and the forms they shape together is what is considered here. In reality, the stars are not at the same distance, and in a 3D projection, the shapes don't hold the same form. The art of identifying forms in the sky and designating constellations is thought to have started several thousand years ago. Related paintings and historical relics are found all over the world. The Greeks adopted the Babylonian constellations around the 5th century BC, and astronomers used these constellations to designate regions in the sky. Nowadays, the sky is divided into 88 constellations that cover the entire celestial sphere. Asterisms are subgroups of stars, within a constellation or shared by several constellations, that observers use to navigate the stars. The constellations crossing the Sun's path are the signs used in astrology. The Sun, as it travels in the celestial sphere to complete a revolution throughout the year, enters 12 apparent constellations that are known as the star signs. Depending on the month you are born in, your astrological sign corresponds to the constellation in which the Sun is located during this month. In fact, there are 13 constellations that the Sun crosses paths with; the 13th constellation is Ophiuchus. However, to match the number of months per year, astrologers decided to adopt 12 signs.

During the astronomical discoveries of the ancient Greeks, other astronomical measurements and discoveries were also made by other astronomers around the world. From those, I mention the Chinese who observed the skies for timekeeping purposes. Chinese astronomy is one of the most precise, with discoveries and documentation of celestial objects dating from 1300 BC. The Chinese developed three models of the Universe; the first one pictures the Earth as flat with an egg-shaped 'heaven' region around it, and the second one sees the 'heavens' as a celestial semi-sphere. In both models, the heavens are filled with vapor surrounding the floating celestial bodies. Earth and the 'heavens' are researchable and observable, and beyond them lies an infinite cosmos. The third model was more of a philosophical notion concentrating on the celestial bodies floating in extent, rather than being a scientific theory. Islamic astronomy, tracking the movements of the Sun and Moon, flowered during the Islamic golden age (8th -13th century), and it played a significant role in reviving European astronomy (after several centuries of historical gaps). In addition, many of the celestial objects nowadays have Arabic names due to the efforts of Arabs in astronomy. Until today, the Islamic calendar and holiday dates follow the Moon calendar, which follows the phases of the Moon in the sky. Islamic astronomy also proposed several alternatives to the geocentric model. Moreover, Arabs are credited for inventing several astronomical instruments before the telescope to help them keep track of the position and movement of stars and to keep track of time.

It is believed that Indian astronomy started in the 15th century. It was influenced by -—and influenced— European, Chinese, and Arab astronomy. Many historical carvings testify that Native Americans were also interested in Astronomy, especially for mythological and timekeeping reasons. African, especially Egyptian, astronomy is also considered important in antiquity. A 7,000-year-old stone circle in the Nubian desert, used to track the Sun, is believed to be the most ancient astronomical site on Earth. These are just a few examples of the history of Astronomy around the world. Astronomy is one of the oldest sciences on Earth, and it is believed that every part of the world has taken an interest in it at a certain point in time. After all, who would not be intrigued by all the wonders above?

Astronomy affects our daily lives, which are centered and organized around the movements of celestial objects, like the Sun and the Moon. Therefore, it is common to find historical indices worldwide of practical ways to track celestial objects. To document it all, I need to dedicate an entire book to that. For this book, however, I will stick to the main historical ideas, events, and people who shaped our understanding of the grand design.

The Emergence of the Heliocentric Model

Although there were many hints that the Earth could not be the center of the Universe in the early ages, the Heliocentric model owes its rise to Nicolaus Copernicus in Renaissance times. The Heliocentric model states that the Sun is the center of the Universe and that Earth rotates around the Sun and revolves around itself. We are getting closer to the truth, but we are still very far. Copernicus worked on heliocentricity for 40 years and did all his observations with the naked eye (Galileo invented the telescope 50 years later). Between the years 1507 and 1515, Copernicus started circulating the theory of heliocentricity. To keep a realistic model, he held the concept of epicycles. He calculated the distance to the other planets using geometry and placed them on orbits to scale. At the time, the known planets were Mercury at 0.38 au, Venus at 0.72 au, Earth at 1.0 au, Mars at 1.52 au, Jupiter at 5.2 au, and Saturn at 9.1 au from the Sun. Uranus and Neptune were not known at the time. You will find in the next chapter that Copernicus's estimations were pretty close to the actual measurements. The heliocentric model places the stars at a much greater distance than the planets, and it assumes that the air is around the Earth and moves with it. The planet's speed of rotation around the Sun decreases with increasing distances. The book where Copernicus postulated his heliocentric model was first printed when he was on his deathbed. The book's preface, which he did not write, states that this is just a model and does not reflect reality. The real fuss came with Galileo many years later.

Tycho Brahe was a Danish astronomer born in 1546 who was very keen on the precision of his astronomical measurements. His measurements and planet tracking were the most precise then, and he developed several naked-eye measurement instruments. Brahe is also considered by many to be the first modern astronomer who relied on empirical facts to build his models. He left a catalog of more than a thousand stars for the astronomical community. Besides losing a part of his nose during a duel with his cousin, he is known for inventing a hybrid cosmological model where the Sun and the Moon orbit the Earth while other planets orbit the Sun. He is also known to be the first to suggest elliptical orbits for comets, and stipulate that these objects are outside Earth's atmosphere.

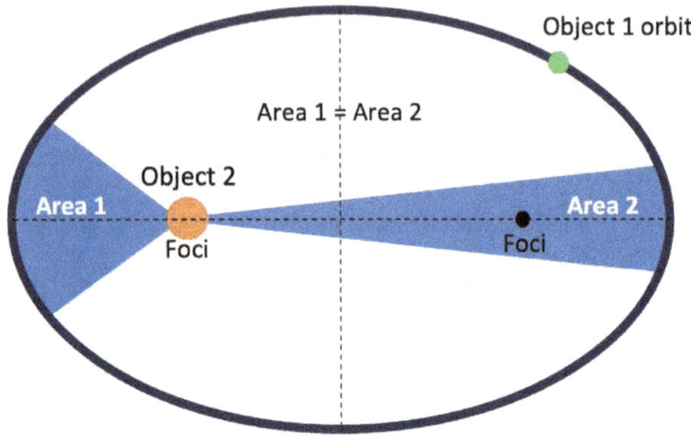

Figure 2.5

Johannes Kepler was a German astronomer and one of the key figures in astronomy and the scientific revolution. Tycho Brahe asked Kepler to prove his calculations in 1600 and Kepler decided to do so by starting his calculations again from scratch. He introduced the elliptical shape of the orbits. He is the father of the ellipse, the planet's orbital shape, and he is known for his three laws. The first law states that the orbits of the planets have an elliptical shape, and the Sun is at one of its foci. The second law states that planets sweep equal areas in the ellipse by equal time (Figure 2.5). The third law is a mathematical relation between the orbital period of the planet and the semi-major axis of the orbital ellipse. Kepler was searching for the reason behind the planet's movements, but he died before finding a convincing answer.

From Galileo to Einstein

Galilco Galilei is the author of *The Dialogue.* He was the first to use a telescope for astronomy, and with the help of his telescope, he made major discoveries. He found that the surface of the Moon was not flat and discovered 'mountains on the Moon' similar to Earth. He discovered the four major moons of Jupiter (the Galilean moons), which indicated that Earth was not the center of all rotations; on the contrary, some objects orbit different objects than the Earth and the Sun. Galileo also observed sunspots proving once more that the Sun (a celestial object) is not perfectly shaped. He also discovered that many parts of the sky contain invisible stars; stars that are not visible to the naked eye. Using his telescope, he observed the surface of the planets. Before telescopes, we could see planets, with the naked eye, only as dots. Galileo discovered that planets were not actually "wanderer stars" like it was thought of for centuries. He observed the phases of Venus (yes, Venus has phases like the Moon), and noticed that during the Crescent phase, Venus looked large, indicating that it was close to Earth, and during the Gibbous phase (when the planet looks like a "croissant", bulging outwards), it looked small, suggesting that it was far away. These observations do not fit a geocentric model but seem more acceptable in the Heliocentric model.

Galileo concluded that the Sun is the center of the Universe and that Earth rotates around the Sun. These discoveries, which diminish the role of Earth in the Universe, unleashed the wrath of the Church. Galileo's book, *The Dialogue*, was prohibited from the public, and he was imprisoned. He continued working on his theories in astronomy and mechanics and died in 1642. Three centuries later, the Church admitted that Galileo was right.

In 1643, right after the death of Galileo, Isaac Newton was born. He is considered by many the greatest scientist of all time. Contrary to popular belief, the apple that helped Newton discover gravity did not fall on his head. Instead, the apple's motion inspired him while it was falling toward the ground. He thought some force was acting on the apple to let it move toward Earth. He also made the assimilation with the Moon and thought that if the Earth did not attract the Moon by some force, then the Moon would move away in a straight line. He called this force: gravity. Convenient! Gravity is a force of attraction that a body with a mass exerts on another body. Newton used his discovery to explain the motion of celestial objects.

Brahe, Kepler, and Galileo described the motion of the planets and shaped the new model, and Newton explained the cause of the motion with his known three laws of motion. The first law states that if an object is isolated, or left alone, it will stay at rest or move in a straight line at a constant speed. The second law is a relation between the force, mass, and acceleration of objects. The third law states that when object X exerts a force on object Y, object Y exerts an equal and opposite force back on X. If a car hits you, you exert on the car the same force that the car exerts on you. That is why the hood gets damaged. Of course, the effects of this force on flesh and bone are much greater than its effects on metal. Newton also invented calculus and made many discoveries in the field of optics. It is worth mentioning that he did most of his work, notably the laws of motion, calculus, and his work on optics in 1665 while he was quarantining during the plague outbreak. He died in 1727, and it is believed that he never had a romantic relationship.

Many scientists came after Newton. In 1781, William Hershel discovered Uranus; the 7th planet in the solar system. He initially thought it was a comet or a star. J.C Adams and U. Le Verrier used Newton's law of motion and worked independently to predict the existence of an 8th planet. They did that by observing Uranus's trajectory and realized that a heavy object was affecting its orbit. In 1846 Neptune was discovered by Johann Gottfried Galle and Heinrich Louis d'Arrest, close to where Adams and Le Verrier predicted it would be. It is the first planet whose existence was predicted before it was actually seen.

Albert Einstein was a theoretical physicist born in 1879. Oh, there are a lot of things to say about Albert. However, to finish this chain of thought, I will summarize his main contributions to science. So, now we know the exact motion of the planets, that the Earth revolves around the Sun, and that the cause of this motion is gravity. But how does gravity work? What is behind this force of attraction? In 1905,

Einstein published four groundbreaking papers, one of which was the Special Theory of Relativity. There are many takeaways from the theory of special relativity, but the most relevant one here is that the laws of mechanics developed by Newton could no longer describe everything in the Universe. For example, one of the many facts Einstein established is that the flow of time is not constant; it is relative to the frame of reference (hence the name relativity) in which the measurements are made. For example, consider two people. One is standing still on the ground while another is sitting in a very high-speed moving train, traveling at speeds close to the speed of light. The second person stays on the train for 5 minutes and then steps out. This person will find that the person standing outside had aged 20 years. Yes, the person on the ground has waited 20 years, while the person inside the train barely had the time to smoke a cigarette (at the time, smoking was allowed indoors). Without going into many details, the notion of time is affected by speed, and it is relative (the same goes for length).

Relativity is the theory that established the speed limit in the Universe: the speed of light. In more detail, everything moves at the maximum speed of the Universe, but the speed components are divided into space and time. Light particles, called photons, have no mass and do not experience time. Light only moves in space; therefore, it moves at the maximum speed in space. Special relativity is the theory that gave the famous $E=mc^2$ equation. $E=mc^2$ states that mass and energy are the same and can be interchangeable. In 1916, Einstein expanded his theory to gravitational fields and discovered that gravity is not a force, but it is an illusion of a force. Let me explain. Einstein had already postulated that space and time are interconnected in the same fabric and that it is a 4D space-time: 3 dimensions of space and 1 dimension of time. He explained how gravity works and how the effect of bending space-time by mass or energy simulates a force of attraction. This is described in the water mattress example given in Chapter 1. General relativity is based on the equivalence principle that states that gravity and acceleration are the same. To be accepted widely, a theory needs to be experimentally proven. General relativity has many implications and these implications can be used to verify the theory. One of them is that light will follow the curvature of space-time caused by a massive object, like the Sun. Therefore, the direction of the light passing near massive objects will change. For example, the positioning of the stars behind the Sun will appear slightly displaced due to this effect; as if the (bent) light rays from the stars are coming from a different place. However, it is impossible to see the stars around the Sun in the daytime due to the brightness of the Sun.

The opportunity to prove general relativity came in 1919 during a total solar eclipse and an expedition organized by Frank Dyson. Sir Arthur Eddington traveled to the West African island of Principe, where the total solar eclipse was visible. During the eclipse, the stars became visible due to the dimming of the sunlight caused by the Moon. Einstein was unknown then, and proving his theory was a rare opportunity. On May 29, 1919, Sir Arthur and his expedition measured the position of the stars during the solar eclipse. The position of the stars had indeed shifted from the original one as postulated in the theory of general relativity. This was the first definite experimental proof of Einstein's general relativity. The results were presented in a joint meeting of the Royal Astronomical Society and the Royal Society. At this moment, Einstein was perceived in high regard, and his fame elevated to the level of a pop culture superstar. Since then, general relativity has been proven in many other ways. For example, due to relativistic effects, the motion of Mercury (the closest planet to the Sun) deviates from the predictions of Newton's laws of motion. Observations showed that Mercury depicted exactly the motion predicted by the theory of general relativity.

General relativity is still being tested nowadays. Nearly a century later, the first direct detection of gravitational waves was made, and it was due to Einstein's general relativity theory. Einstein won the Nobel Prize in physics in 1921, not for his theory of relativity, but for his discovery of the Law of the Photoelectric Effect published in one of his papers in 1904/1905. The general theory of relativity was very controversial at the time, and it was not widely accepted until many years later.

So far, we've mentioned that the Earth and other planets rotate around the Sun, that the cause of this motion is gravity, and we understand how gravity works. So, let's consider further thoughts. Is the Sun the center of the entire Universe? What about the shape of the Universe, the size of it, its beginning, and its end?

Astronomy Today

At the time of Einstein, astronomers and scientists knew that the Universe was much bigger than they had thought and that it might be infinite and full of stars. The discovery of new stars started to shape the galaxy. Scientists did not know at the time that the Universe was not only bound to our galaxy, but they started realizing that we might not be in the center of the galaxy/Universe after all. There are some questions that we did not discuss in the previous sections. Is the Universe infinite? Is it moving or static? Is it an eternal Universe, or does it have an ending and a beginning?

Newton argued that if the Universe was finite, gravity would attract everything to the center. But if the Universe is infinite, then there would be no center. Heinrich Olbers argued against the infinity of the Universe, stating that if the Universe was infinite, then the sky should be rather more luminous than dark because every part of it would contain stars. In that case, the lack of a luminous night sky would be explained by the fact that stars have not been shining forever. This means that the Universe is not infinite, or that stars are not eternal.

Einstein was a believer in a static infinite Universe at first. He also encountered a similar paradox in his theory. In his general theory of relativity, in the equation describing the Universe, he found that to maintain a static and infinite Universe, he had to add a constant to his equation: Einstein's Cosmological Constant. Spoiler alert: He later admitted that it was his biggest blunder, especially since he failed to trust the story his equations told. He then discarded the constant.

Edwin Hubble is an English scientist with a powerful telescope. Around 1924 he discovered that nebulas (bodies in the sky that look like clouds) contain stars. He realized that those were other galaxies and not bodies in our galaxy. The Universe just got much, much bigger. At the time Visto M. Slipher discovered shift of spectral lines of distant galaxies (redshift). Hubble also discovered that distant galaxies are moving away from us, which is proof that the Universe is not static. When a body moves away from us, its color shifts to red. The more distant the galaxy, the more the color shifts to red, which also means it is moving away faster. This shows that the Universe is expanding at an increasing level. Independently, Georges Lemaître also discovered the expansion of the Universe. This gave the Hubble-Lemaître's law (by Hubble and Humason) that establishes a proportional relation between the distance and the speed of a galaxy.

Sixty years later, the most powerful telescope, taking Edwin Hubble's name, was sent to space and changed the way we perceive the Universe. One of the most impressive discoveries that the Hubble telescope made is the Hubble Deep Field. The telescope was pointed toward what was considered an empty region and discovered thousands of galaxies in that region. Nowadays, we estimate that the Universe contains trillions of galaxies. The most accepted theory about the beginning of the Universe is the Big Bang theory. The Universe, space, time, and everything in it started with a small point called a singularity that inflated and started expanding. The Universe is expanding, but what is important to know is that it is space-time in itself that is expanding, and everything in it is undergoing this expansion. What is NOT true is imagining the Universe as a static room where things were compressed in a small point at the center and they started moving away from each other. It would be more accurate to say that the room itself is inflating. In the 1990s, scientists discovered that not only is the Universe expanding, but that the rate of acceleration is also increasing; it is an accelerated expansion. I will discuss the notions of cosmology later in the book. Small spoiler: Everything that we see in the Universe, all the regular matter, the galaxies, stars, planets, etc. only constitute around 5% of the actual Universe. There is around 25% of dark matter and about 70% of dark energy in the Universe. We do not understand 95% of the Universe. Dark energy is believed to be behind the accelerated expansion. Ironically, Einstein's cosmological constant can be used today to account for this dark energy. He was not entirely wrong to add it after all.

Chapter 3

The Solar System

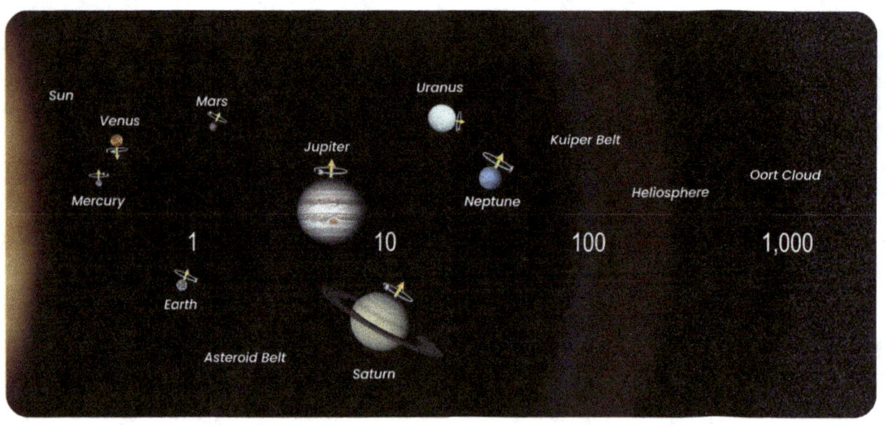

Sun
Venus
Mars
Uranus
Mercury
Jupiter
Kuiper Belt
Neptune
Heliosphere
Oort Cloud
1
10
100
1,000
Earth
Asteroid Belt
Saturn

The solar system consists of the Sun, planets like Earth, moons like the Moon, and other small objects such as dwarf planets, asteroids, comets, dust, and gas. The solar system will remain humanity's home for now and in the near future. This chapter will help you understand the system's main components. It all started with clouds of gas and dust. Due to their gravity, the particles of dust and gas were attracted to each other, and the cloud collapsed and formed the Sun. However, not all the particles participated in the formation of the Sun. Other residues stayed out there and created the remaining solar system objects. Let's take a deeper look into the Sun and these other objects.

The Sun

The Sun is a star. Like all stars, it is mainly formed of hydrogen and helium, along with other gasses and a little bit of heavier elements — usually smaller than a grain of sand —referred to as dust. But how do we get light and heat from a body of gas? Well, let me tell you. As explained earlier, everything in the Universe that has a mass also has gravity. It all started as a cloud of gas and dust. The gas and dust came from older stars that died, but we will get to that later. The particles in the cloud are attracted to each other by gravity, and they start to collide creating friction and heat; the heat increases the particle movements, and the increased movement creates more heat. The heat increases, reaching several million degrees in temperature, and with those temperatures, nuclear fusion takes place. In a nutshell, two hydrogen atoms will collide and fuse into one helium atom with a mass equivalent to the two hydrogen atoms. Well, not exactly! The atom of helium has a slightly lower mass than the mass of the two atoms of hydrogen; the rest is dissipated into particles and electromagnetic radiation (the source of light and heat). This is how the ignition starts, and it remains that way for billions of years. When electromagnetic radiation is created by the fusion process in the core of the Sun, the rays are created in a very dense medium. The light will be absorbed and re-emitted several times, significantly increasing the time it takes to reach the star's surface. It can take light 100,000 years to escape the core and get to the surface before it escapes the star.

The Sun first becomes a protostar, with a disc of matter rotating around the core. The matter is accreted until a sphere is formed. The Sun has then matured from a cloud of gas and dust to a ball of fire; not real fire, but a ball of nuclear reactions. This all happened 5 billion years ago. The Sun's life expectancy is directly related to its mass. The heavier a star is, the faster it burns its reserves, and the quicker it will die. The Sun is expected to have a lifespan of 10 billion years, which means that the Sun will die in 5 billion years. The Sun has a radius of around 700, 000 km. In comparison, Earth has a radius of around 6,400 km. The Sun is much bigger than Earth. If the Sun was empty, around a million Earths could fit inside (do not confuse radius, surface, and volume). If the Sun was a football, then Earth would be smaller than the tip of a needle in comparison. This is how you can imagine it.

The Sun is the center of the solar system and is by far the heaviest and the most prominent object in it. It has a mass of ~2 x 10^30 kg (2 with 30 zeros). However, compared to other stars, the Sun is relatively small; but more on that later. The Sun's surface temperature is around 6,000 degrees Celsius (°C), and its core temperature is 15 million °C. The Sun comprises a very dense and hot core, with two layers englobing it. The two layers differ in how they transfer energy; one through radiation and one through convection. The convection zone extends to the surface of the Sun, the Photosphere. After the photosphere is the Chromosphere, a thin layer of plasma, and the outermost region of the

Sun is the Corona; it can be thought of as the atmosphere of the Sun. The Corona is the part you can see from the Sun during a total solar eclipse and its temperature ranges between 1 to 2 million °C. The Sun's atmosphere is much less dense than the Earth's atmosphere. Small frequent solar flares might be the reason behind the hot temperature of the Corona. The Sun has a vast magnetic field that is twice as stronger as Earth's but extends to farther regions. The magnetic field extends to the solar system's outer edge (you will know how large the solar system is at the end of this chapter). Looking closely at the Sun with a special telescope, we see black patches, called Sunspots, and they are cooler regions on the surface of the Sun (~4,000°C). These spots are caused by an intense magnetic field that inhibits gas escape from the Sun's interior layers.

As explained before, the Sun is a ball of gas that changes size when it ages. At the beginning of its formation, the Sun was larger than it is today, but it has shrunk with time. As it reaches the end of its life, in 5 billion years, when all the hydrogen reservoirs in the core are depleted, the Sun will begin fusing hydrogen in the outer layers and will start swelling. It will become a red giant (literally). It will continue to swell, getting closer to the inner planets: Mercury, Venus, and Earth, potentially engulfing (and roasting) some of them. At this stage, the Sun will continue to swell, and eventually fade into a cloud of gas and dust. A dense core will remain, however, and this dead star is what we call a white dwarf. I will explain white dwarfs in the next chapter, but they are balls of highly dense matter that can stay in this state for billions and billions of years.

Around each star exists a zone with an ideal temperature for liquid water. This zone is called the habitable zone. If a planet is very close to a star, like Mercury or Venus, it will be too hot to have liquid water on the surface, and if there is water, it will be in the form of vapor. If a planet is too far, then there is not enough heat reaching it from the star, and it will be too cold. So, if water exists on its surface, it will be in a frozen form. If a planet is in a habitable zone, not too far and not too close, water can exist on the surface in liquid form. The habitable zone is also called the Goldilocks zone (from the fairy tale). Earth, today, is in the Sun's habitable zone. An example is given in Figure 3.1.

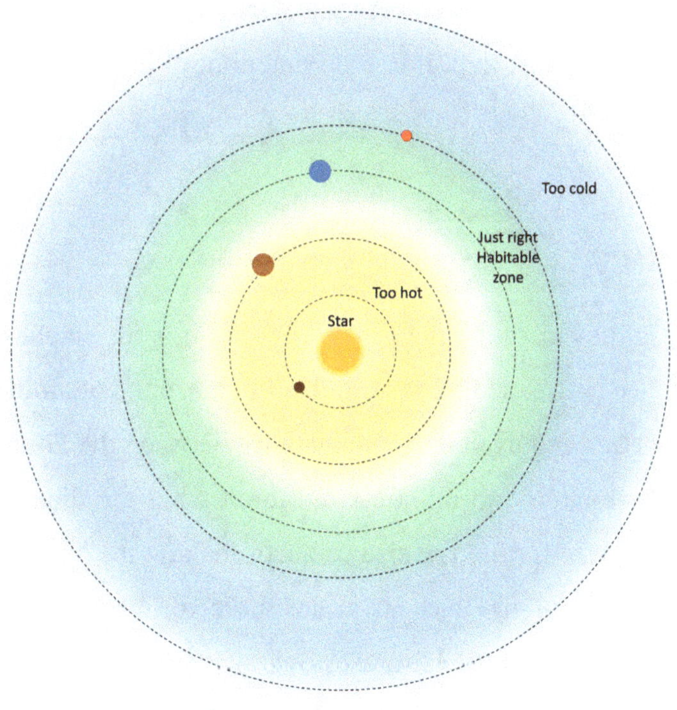

Figure 3.1

When the Sun was accreting matter while forming, a portion of matter did not fall into the Sun. It stayed gravitationally bound to it and kept rotating around it. 5 billion years ago, the solar system was very dense. Far away from the extreme heat of the Sun, matter started condensing. This matter around the Sun began to collide and merge. Small rocks started forming from the accumulation of matter. Small rocks collided and combined to form bigger rocks. Bigger rocks collided and merged to form planetesimals, and then planetesimals formed planets. Not all rocks merged into planets. There are two types of planets in the solar system: rocky planets that have rocky surfaces, and gaseous planets that are made of gas and have rocky cores. In the solar system, rocky planets are the inner planets Mercury, Venus, Earth, and Mars. The gaseous planets are the outer planets Jupiter, Saturn, Uranus, and Neptune. But what about Pluto?

Mercury

Mercury is the closest planet to the Sun and it is the smallest planet. The distance between Mercury and the Sun is around 68 million km or 0.387 au. 1 au is equivalent to the distance between the Earth and the Sun, approximately 150 million km. Mercury has a radius of ~2,400 km; its total size is 38% the size of Earth making it closer to the size of the Moon. Mercury has no atmosphere; probably because it is close to the Sun, and the latter's strong gravity swallowed Mercury's atmosphere. Mercury has extreme temperatures: the side facing the

Sun has a temperature that can reach around 430°C, and the opposite side (the night side) has a temperature that can drop to around -180°C. Mercury's revolution period around the Sun is 88 days, meaning it will complete one full orbit, or one Mercury-year, in 88 days (on Earth, a year is 365.25 days). Mercury rotates around itself every ~59 days. One day on Mercury is equivalent to 59 days on Earth. Mercury's surface is full of craters because of its lack of atmosphere which should have contributed to protecting the planet from heavy asteroid bombardment. Three missions have been sent to the planet that aided the discovery of several features like a magnetic field and a surface full of craters. These missions studied the planet's surface, geological history, and magnetic field. BepiColombo was the latest mission sent in 2018, and it is still ongoing. Amongst other things, the mission confirmed that the poles of Mercury contain ice (water). It turned out that ice is more common on the planet than previously thought.

Venus

Venus is the second closest planet to the Sun. The distance between Venus and the Sun is approximately 0.72 au. Its size is comparable to Earth, with a radius of ~6,000 km. Venus is the hottest planet in the solar system. You might wonder why Venus and not Mercury. Venus has a very dense atmosphere composed mainly of carbon dioxide and sulfuric acid; it is not a friendly environment. This results in a very strong greenhouse effect on the surface of Venus. The Sun's rays enter the dense atmosphere and are trapped by the atmosphere leading to an astonishing temperature of ~464°C. The atmospheric pressure on the surface is more than 90 times stronger than on Earth. Standing on the surface of Venus is similar to being under 900 meters of water; you would be crushed. Venus revolves around the Sun once every ~225 days and rotates around itself every 243 days, making one day on Venus longer than one year. Moreover, Venus is the only planet spinning around itself in the opposite direction; at least, this is how it looks to us today. Scientists believe the planet was spinning in the same direction as the others, but because of its heavy atmosphere, the Sun exerted a strong gravitational pull and flipped the planet upside down, with the help of some complex dynamics of the planet's layers. In this case, the planet spins in the right direction, but upside down. Venus also has a particularity that makes it the brightest object in the night sky after the Moon; its surface has a very high reflective index. In other words, its whitish color reflects the Sun's light very brightly. Further particularities of Venus show that its surface is full of volcanoes, and most of it is covered by volcanic basalt (rock). Many missions have been sent to Venus; since 1973, 22 probes have visited the planet including several landers. The missions studied the atmosphere and the surface of the planet. Geologically, the planet has no known systemic activity. It has a flat surface and, as mentioned before, is covered with volcanos. It is also challenging to access because of its very thick atmosphere. The planet's main feature is its atmosphere which is, in fact, a learning mine for astronomers and climate scientists. It has been indicated that Venus has no magnetic field.

Earth

Earth is rocky and sphere-shaped, and it is the third planet from the Sun. Earth is around 150 million km from the Sun (~1 au), with a sphere of 12,742 km in diameter, and a mass of ~6 x 10^24 kg (6 with 24 zeros or 6 trillion, trillion kg). The planet's average temperature is around 13°C, with record values reaching around -90°C in the coldest regions, and around 58°C in the hottest regions; in extreme seasons and extreme cases. The atmosphere is composed of 78% nitrogen, 21% oxygen, and 1% other gases. Earth revolves around the Sun once every 365 days, 6 hours, and 9 minutes. It travels a distance of ~30 km/s in the solar system around the Sun. One year on Earth is counted as 365 days, except for leap years which come once every 4 years and have 366 days. On average, it takes Earth 365.25 days to complete a full orbit around the Sun. The added day accounts for the 6 extra hours each year that Earth needs to complete a full orbit around the Sun.

Earth rotates around itself; this is why we see the Sun rise and set at the horizon. Earth completes a rotation around itself once every 24 hours, which determines the length of a day. The actual value of Earth's rotation period is 23 hours and 56 minutes. Since Earth is simultaneously moving around the Sun, it takes four extra minutes to return to the same position relative to the Sun, as shown in Figure 3.2. The Earth's rotation speed is around 1,600 km/h at the equator.

We don't feel the movement of the Earth because it is large. The speed is also relatively constant; we can only feel accelerations and speed changes. Moving at a constant speed and being at rest ultimately have the same effect. Think of it, when you are in a car, no matter how fast it is moving, you will sit comfortably in your seat. It is only when the car accelerates or decelerates that you feel it. Earth is tilted on its axis forming an angle of 23.5° with the axis of the solar system's plane.

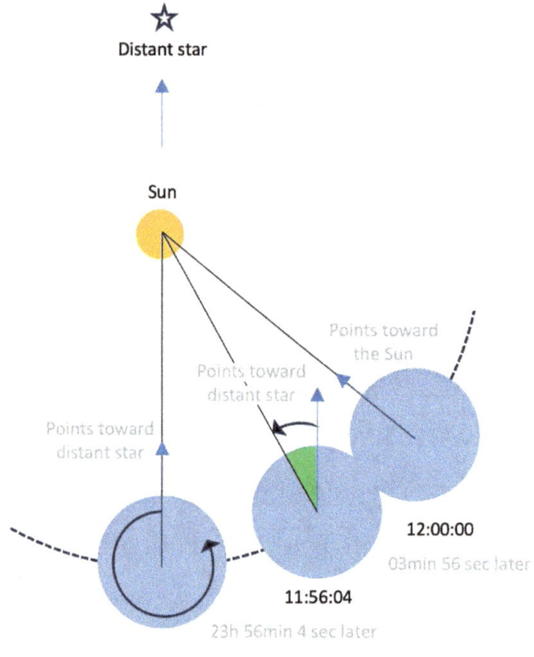

Figure 3.2

Reiterating some of the ideas from Chapter 2, the projection of the Earth's equator on the celestial sphere is called the celestial equator, and the projection of the solar system's plane is the ecliptic, as shown in Figure 3.3. In other words, the ecliptic is defined by the position of the Sun every day at noon, over the year.

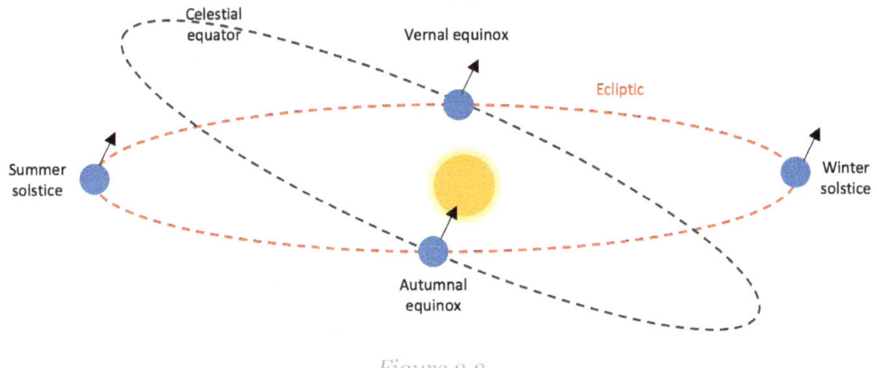

Figure 3.3

Due to Earth's tilt, the celestial equator is not aligned with the solar system's plane. The seasons on Earth are due to its tilt and not its distance. While Earth revolves around the Sun, the position of the sun changes in relation to the celestial equator. In winter, the Sun at noon is below the celestial equator, and in summer it is above it. The elevation of the Sun in the sky makes the temperature rise and fall; when the Sun gets higher, the temperature rises. The Sun is lowest at the beginning of winter and highest at the beginning of summer; these are called the summer and winter solstices. This also defines the longest and shortest days of the year, respectively. The northern and southern hemispheres experience summer and winter in opposite ways. When it is summer in the north, for instance, it is winter in the south. There are two days in a year when the Sun at noon is at the level of the celestial equator. It is at the intersection of the ecliptic and the celestial equator. These days are the beginnings of spring and fall, around the 21st of March (vernal equinox) and September (autumnal equinox) respectively. The positions of objects in the sky are calculated with respect to the celestial equator and the vernal equinox. Around

70% of Earth's surface is covered by water bodies. Three states of water can be found on Earth: gas in the atmosphere, liquid in oceans, seas, lakes, aquifers, and rivers, and solid in the poles and cold regions. The highest mountain from the surface of the ocean is Mount Everest which stands at 8,849 m from the surface. However, Mount Everest is not the highest peak if we measure the distance from the center of the Earth. Due to Earth's rotation, it slightly bulges at the equator, which makes the distance from the center to the poles slightly smaller than the distance from the center to the equator. Mount Chimborazo's peak, in Central Ecuador near the equator, is the farthest point on the surface from the center.

Earth appears blue from space. Our sky is blue, and the color reflects on the transparent bodies of water. Sunlight is composed of a spectrum of colors, known as the rainbow colors. The sky is blue because blue light is the most energetic in the visible spectrum, and is easily scattered everywhere by molecules in our atmosphere. As the Sun sets, the light coming from the Sun has to travel a larger distance in our atmosphere. The blue color will be scattered even further, and the larger distance allows other less energetic colors, like red, to scatter and to light up the atmosphere as illustrated in Figure 3.4. This is why the sky turns reddish toward sunset and sunrise.

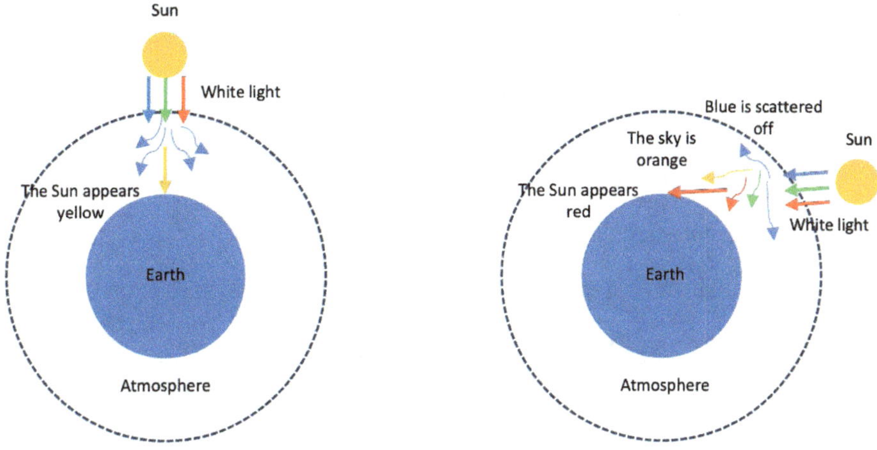

Figure 3.4

Earth has a particularity that makes it unique: it has life; it is our home. Many characteristics of the planet came together to make this possible. Of which: Earth's position in the habitable zone of the solar system, it contains liquid water on its surface and oxygen in the atmosphere, it has a strong magnetic field, and it has an atmosphere that protects life from lethal cosmic radiation. These characteristics, in addition to many more, make it possible to sustain life on our planet.

The Earth formed 5 billion years ago while the solar system was still in its youth. It was initially a red-hot planet that cooled with time. Earth has one natural satellite; the Moon. It is believed that the Moon formed in the early stages of the solar system when countless asteroids and small bodies heavily bombarded Earth. It is thought that a body nearly half the size of Earth (known as Thea) collided with Earth and threw enough debris that combined and formed the Moon. The Moon is around a quarter the size of Earth (in diameter). It has no atmosphere, and its surface is full of craters created by heavy asteroid bombardment. The Moon and the Sun exercise gravitational pulls on Earth, and this creates tides on the two sides of the planet. These tides pull the water from both sides, which makes it rise above its normal level. While rotating, the land enters regions where water is pulled, and we experience high rises in ocean levels. A visual description of this phenomenon can be seen in Figure 3.5.

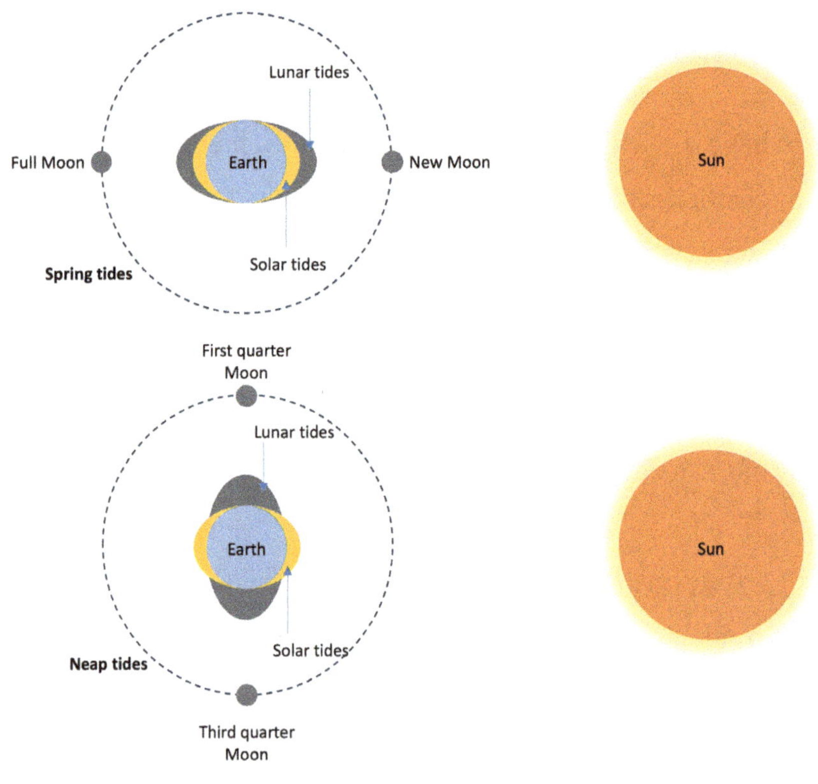

Figure 3.5

The Moon's revolution around the Earth and its rotation around itself are synchronized. This means that we always see one side of the Moon.

Relative to Earth, the Moon has a near side that always faces it, and a far side that is never seen from the planet's surface. To our knowledge, the Moon is the only other solar system object that has hosted human life on its surface for very brief periods. This happened when humans visited the Moon in the 1960s and 1970s. Many artificial objects, such as probes and rovers, sent by different countries, in particular the USA and the USSR, visited the Moon. All rovers that were sent, landed on the near side of the Moon. One of the last missions to land on the surface of the Moon was the Chinese Chang'e 4 spacecraft in 2019. Chang'e 4 landed on the far side of the Moon and was the first spacecraft to do so. Very recently, India landed a probe and lander, Vikram and Pragyan, on the uncharted south pole of the Moon. I discuss crewed missions to the Moon in Chapter 10.

Mars

Mars is the most explored planet by humanity aside from Earth. Its average distance to the Sun is 1.52 au, and it is a relatively small planet with a radius of 3,390 km, which is a bit more than half the radius of Earth. Its atmosphere is composed mainly of carbon dioxide, so it is unsuitable for life now. The temperature on Mars's surface can drop to -100°C in certain regions during the nighttime, and rise to 20°C during the day. Its surface is subject to important temperature variations throughout the day and through different seasons, with an average of -65°C. Mars completes a full revolution around the Sun in 687 days (almost two years on Earth), and a complete rotation around itself in 24.6 hours; close to one day on Earth. Mars is often thought to be our next destination and our next home. With current technology, getting to the planet would take around 6 months. With interplanetary voyages, the trips are not straightforward or in straight lines. Many factors must be considered, like waiting for Earth and Mars to get closer, since they both asynchronously orbit the Sun, and can be on opposite sides. Moreover, the trip would have to include many gravitational assist flybys, such as gravitational slingshots, to reach the desired travel speed with the least energy consumption. A gravitational slingshot is the process of flying by a planet (or other objects) and using its gravity to boost a probe's speed.

In addition, Earth and Mars are constantly in motion, which is another factor that must be considered. The long duration of the trip is one of the main limitations of interplanetary human travel. Spending such a long time in an aircraft can be unbearable; try sitting in your room for 6 months straight without leaving it. Mars cannot support human life, so people living there must stay in controlled environments, like domes, that are adapted for extreme temperature variations and contain breathable air. The Martians would also have to be cautious of dust storms, which are very frequent on Mars. They also have to get used to the weak gravity as it is only a third of the gravity on Earth's surface. Weak gravity affects human physiology because our bones and muscles are adapted to 3 times the force. But what about newborns on Mars? If they grow up in weak gravity and their strength, bones and muscles get used to it, will they get crushed if they return to Earth? They would probably have to exercise before their trip to the mother planet. Mars has two small moons, Phobos and Deimos that are a few kilometers in size – very small compared to the Moon. Olympus Mons is a volcano on the surface of Mars that occupies the spot of the tallest mountain in the solar system, rising at 25 km above the surface; it dwarfs Mount Everest.

Many human-made probes visited Mars; one of the latest ones is the Emirates Mars Mission: the Hope probe. The probe is studying the planet's atmosphere, climate, and weather. This probe is the first mission to another planet that was carried out by an Arab country. Mars is the only planet with human-made rovers on its surface as we speak. Five rovers have examined the surface, of which two are still active: Curiosity and Perseverance. The latter reached Mars in 2021. These five rovers, Curiosity in particular, found hints of ancient streams of water on the planet. The planet contains water but mostly in ice form, and some in vapor form. Moreover, scientific evidence suggests that Mars had weather conditions suitable for life a few billion years ago. It is even suggested that liquid water was running on the surface of the planet and that it was in the habitable zone of the Sun before the latter shifted toward Earth's orbit.

Asteroids and the Asteroid Belt

The asteroid belt is a region in the solar system between the planets Mars and Jupiter that is full of asteroids. An asteroid is a rocky object that orbits the Sun. It is believed that not all the rocks and planetesimals merged to form a planet, but that some remained as asteroids. One of the leading causes that prevents asteroids from merging into a planet is Jupiter's strong gravitational perturbations. Asteroids come in different shapes and sizes, varying from a few meters to several hundreds of kilometers in diameter. Asteroids are everywhere in the solar system, with an important concentration around the belt. Other types of asteroids are the near-Earth asteroids and the Trojan asteroids. Trojan Asteroids are asteroids that share the orbit of some planets due to gravitational effects. Most Trojan asteroids are found in the orbit of Jupiter.

The largest asteroid in the asteroid belt is Vesta (with a diameter of 525 km), but it is not the largest object in the belt. The largest object is Ceres, the first object discovered in the belt. However, Ceres is classified as a dwarf planet due to its spherical shape and much larger size. Ceres has a diameter of 946 km but is still much smaller than an actual planet. Ceres accounts for a quarter of the mass in the asteroid belt; the entire asteroid belt is less than 4% the mass of the Moon.

Asteroids mainly contain rocks and metals, and some contain water. The more metal (iron in particular) an asteroid contains, the more solid it is. Some smaller asteroids orbit around other larger asteroids, while other asteroids are composed of several small rocks bound together by gravity. During their voyages to the outer planets, several probes passed by the asteroid belt. Dawn was the only mission explicitly designed to study the asteroid belt, and it was sent to orbit two of the largest bodies in the belt, Ceres and Vesta. Two other missions were sent to study near-Earth asteroids; one of them is the Hayabusa mission, a Japanese probe sent to collect samples and return them to Earth.

Some asteroids can pose a danger to life on Earth. After all, it is believed that an asteroid wiped out the dinosaurs. When a small rock (or grain) falls on Earth it will burn in the atmosphere under the effect of friction with the air and light up. These small rocks are called *meteoroids*, but when they totally disintegrate in the air, we call them *meteors*. If they survive the trip to the ground, we call them *meteorites*; a harmless, friendly firework show, often referred to as shooting stars. However, the bigger the piece of rock is, the more dangerous it becomes.

Asteroids are hazardous because of the kinetic energy they carry. Imagine an asteroid the size of a school bus heading toward Earth. A school bus made of rocks would weigh 360 tons. Let's assume that this is the mass of the asteroid, 360 tons. The asteroid hits Earth at a speed of 30 km/s. The kinetic energy of the asteroid is around 10^{14} joules. This energy would be released upon impact and is around 10 times more energetic than the Hiroshima nuclear bomb. Now, imagine that the asteroid is a few kilometers wide... that is an extinction-level event. The kinetic energy is proportional to the asteroid's mass, so the bigger the asteroid, the bigger the impact.

The speed of the asteroid plays a major role here as the kinetic energy is proportional to the square of the velocity. Other than the massive local damage that an asteroid would cause, the impact will also cause earthquakes, tsunamis, dust and rock ejecta (that could remain in the atmosphere for years), as well as climate and weather changes. Today, most asteroids are monitored, and their danger level is assessed. No severe threats are identified in the near future. For some time, Apophis, a ~340 m near-Earth object, was on the risk list, but new observations ruled out any impact risks as the asteroid got closer. This does not, however, rule out that our monitoring might not miss other dangerous asteroids.

In case of a direct threat, what do you think we should do? One option would be to nuke the asteroid, but breaking the asteroid into a million little pieces might not solve the problem. After all, these pieces will still fall on Earth. The best way to avoid an impact is by recognizing the threat ahead of time and deviating the asteroid from its course. One way of doing that is to launch a heavy projectile toward the asteroid to shift it from its path. Double Asteroid Redirection Test (DART), was the first attempt to do that. The DART mission consisted of sending a heavy projectile to impact the harmless asteroid Dimorphos, orbiting a larger asteroid, Didymos. Evidently, asteroids can orbit other asteroids as they could be gravitationally bound to them. The results from this first test came back victorious, as the impact indeed deviated the asteroid from its initial trajectory.

Now let's discuss gaseous planets. First of all, how do we get gaseous planets? Gaseous planets started as rocky planets, except that these planets grew larger and more massive than the previous four rocky planets. They grew in a region of the solar system away from the Sun, which is relatively cooler in temperature. In this region, the cooler temperature allowed the condensation of some gases, ice, and rocky material, and as a result, many icy bodies grew in these regions. When the planets in this region reached several times the mass of the Earth, their gravity grew stronger, and they started attracting more and more gas, mainly hydrogen (which is the most abundant). It is also believed that these giant planets may have absorbed small planets. The stronger the gravity, the more gas planets attract and the thicker the layer of gas they form, until they appear dominantly gaseous. This is how gaseous planets are formed.

Jupiter

Jupiter is the largest and most massive planet in our solar system. The distance between Jupiter and the Sun is 5.2 au, which means it is 5 times farther from the Sun than Earth. The planet has a radius of ~70,000 km, making it 11 times wider and 318 times more massive than Earth. Jupiter's mass alone is 2.5 times the mass of all the remaining planets in the solar system. The planet is so massive that it does not orbit the Sun, instead, it orbits a center of gravity located slightly outside the Sun. It is also so large that you could fit more than 1,000 Earths inside Jupiter if it were hollow. You could also fit around 1,000 Jupiter planets in the Sun, or more than 1,000,000 Earths in the Sun. The gravity on the surface of Jupiter is 2.5 times greater than that on Earth. One would expect gravity to be much stronger, but the surface of Jupiter is so distant from its core, that its gravity becomes weaker with distance. However, as you go deeper into the gaseous layers of the planet, and closer to the core, the gravitational force increases significantly.

How do you define the surface of a gaseous planet? For gaseous planets, there is no solid surface that you can step on. The concept of a surface for gaseous planets is defined as the point where the atmospheric pressure equals that of Earth's surface, typically set at one bar. Because of its large mass, Jupiter has taken the role of the protector of the solar system. It has an influential gravity that attracts asteroids and comets, protecting Earth from devastating impacts. Jupiter orbits the Sun once every 12 Earth years, but it rotates around itself once every ~10 hours, completing the cycle of one Jupiter day. The planet rotates quickly around itself and has strong winds on its surface that create many swirling storms. If you look at Jupiter with a telescope, you will first notice a big red spot south of its equator. This spot is a storm with winds reaching 650 km/h. The storm's size is comparable to the size of Earth, and it has been active for at least 150 years (some even claim it has been there since the first telescopic observations of the planet in the 1600s). Observations of the spot show that it has been shrinking with time, and it could disappear in a few dozen years.

In total, Jupiter has between 80 and 95 moons. Among these, there are four moons known as the Galilean moons, which are the largest and the most famous: Io, Calisto, Europa, and Ganymede. In a nutshell, Ganymede is the largest of the Jupiter moons, and the largest moon in the solar system. Calisto, the second-largest moon among the four, has an icy surface. In contrast, Io has an entire volcanic surface. Europa has a rocky mantle with salty ocean water underneath, which makes it one of the most interesting moons in the solar system due to its possibility to suit life in its oceans. But how can a moon this far host life? Isn't it too cold? Yes, it is too cold out there, but the sun is not the only source of heat in the solar system. In this case, Jupiter's strong gravitational pull on the moon contributes to its heating.

With potential warmth and salty water oceans, Europa has triggered the interest of space explorers. Several probes have visited Europa while visiting the gaseous giant, Jupiter. Other missions propose to drill into the moon's surface to explore its oceans. In total, nine missions have visited Jupiter until now. The most recent mission is Juno, a probe launched in 2011 and is planned to orbit Jupiter until 2025. Juno reached Jupiter in 2016, after a five-year journey. The next mission is the European JUICE, launched in 2023 and currently en route to the giant planet.

One last note about Jupiter is that it is often called a failed star. It is said that had the planet been 80 times more massive, it could have started a fusion reaction and become a star. This indicates that the minimum mass for a star to form is around 80 times the mass of Jupiter, or 0.08 times the mass of the Sun. But 80 times more mass seems a lot. It is as if Jupiter was ever close to becoming a star. Jupiter would actually need to be more than 10 times more massive to be considered a "brown dwarf"; let alone a star. A brown dwarf is a planet-like object. It is considered a star candidate because it does not have enough mass to become a star; its mass is usually between ~10 and 80 Jupiters. A brown dwarf is a failed star. More information on brown dwarfs will be covered in the next chapter. To me, Jupiter is closer to a failed 'failed star'.

Saturn

Saturn is known as the jewel of the solar system. Why? Because of its rings. Saturn has the most extensive rings in the solar system. Several planets have rings, like Jupiter, Uranus, and Neptune, but Saturn's rings are much bigger. The rings are the first thing you see when you look at Saturn with a telescope. They are made of chunks of rock and ice that can vary in size from as big as a building, to as small as a grain of sand. Saturn appears to have seven main rings among other minor ones. The rings are believed to be chunks of asteroids and comets that have shattered around the planet. Saturn is the second largest planet in the solar system after Jupiter, with a radius of 58,232 km. The planet's rings extend to nearly 280,000 km from the planet. The rings of Saturn are very thin, with just a few meters in thickness. Saturn is 95 times more massive than Earth, but it has the lowest density in the solar system (density = 0.69); lower than water (density = 1). 1 year on Saturn equals 29.4 Earth years, and 1 day on Saturn equals 10.7 hours.

Another feature of Saturn is that it has the most moons in the solar system, with 146 moons discovered until now. Titan is the largest of these moons and is the second-largest moon in the solar system. Titan is the only solar system moon with a dense atmosphere, and the only one with liquid seas and lakes on its surface, mainly composed of methane.

Four missions have visited Saturn in total. Cassini–Huygens is a space probe that was sent in 1997 to orbit Saturn and study the entire system including the planet, its moons, and its rings. The probe arrived to Saturn in 2004 after a seven-year trip. The probe carried another probe on board, Huygens, that was sent to the surface of Titan. Huygens performed the most distant landing known to this date. Amongst other exciting things, Cassini discovered in one of Saturn's moons, Enceladus, plumes of water ejected from its southern pole. This hints at the existence of a liquid water ocean beneath the moon's icy surface. Since Saturn is very far from the Sun, it is cold with an average temperature on the surface being -178°C, and -201°C on Enceladus. But, the gravitational pull of Saturn on the moon can heat its core (like in the case of Jupiter and Europa), which makes it possible to have liquid oceans under the icy surface.

Future missions are proposed to fly by the water plumes of Enceladus to search for signs of life in the water jets. This would be much easier than drilling to reach the oceans.

Uranus

Uranus is the first planet to be discovered with a telescope; Astronomer William Herschel discovered it in 1781. I remind you that the previous planets described here were already visible in the sky by the naked eye, but until a certain point in history, humanity did not know they were planets. While sweeping the sky with his own-built telescope, Herschel discovered the planet Uranus. After several observations, he noticed that this particular "star" orbited the Sun, and he confirmed that it was the 7th planet in the solar system. Uranus is at 19.8 au from the Sun. It completes a full orbit around the Sun every 84 years, and a rotation around itself every 17 hours. The planet is around 15 times more massive than Earth with a radius that is 4 times larger than Earth's radius. In addition, it is 63 times more voluminous than Earth.

Uranus is the only planet tipped on its side and appears to be rolling around the Sun. You can imagine that the planet's poles are where the equator is supposed to be. It is spinning in a clockwise direction, while all the other planets spin in an anti-clockwise direction. Why is it this way? Astronomers believe that in the past, a heavy body collided with Uranus and knocked it to its side. Others thought that it was due to the dynamics of its ring system. Indeed, the planet has a ring system that was much larger in the past. Although they are much smaller than Saturn's rings, they are interesting because they follow the planet's spin, and have a vertical direction perpendicular to the other ring systems.

Uranus has 27 known moons, one of them, Miranda, is said to resemble Frankenstein's monster, (although I don't see it) as if it was assembled together from parts. Unfortunately, there have not been any dedicated independent missions to Uranus. Only one probe has ever visited the planet, Voyager 2, while it was visiting the four outer planets. Uranus is around 20 times more distant from the Sun than Earth, and it took

Voyager 2 nearly 4.5 years to travel from Saturn to Uranus. I am dedicating an entire section to the Voyager missions in Chapter 10.

.
.
.
●

Neptune

Neptune is my favorite planet because of its appealing blue color. Neptune is the 8th, and until now, the farthest known planet in the solar system. The planet is 30 times farther from the Sun than Earth. Neptune is 17 times more massive than Earth, and 4 times wider (close to Uranus's size). Since it is so far from the Sun, the temperature is an astonishing -200°C.

Neptune's discovery is an interesting story. It is the first planet to be discovered using math before it was actually seen. By observing the motion of Uranus, two astronomers, Urbain Le Verrier and John Couch Adams, predicted that there was an 8th planet somewhere affecting Uranus's motion. Johann Gottfried Galle, a German astronomer, then used these calculations and identified Neptune in 1846 close to where Urbain Le Verrier and John Couch Adams predicted it would be. Some people say that the first one to observe Neptune was Galileo Galilei while observing Jupiter, but he mistook it for a star because it was moving very slowly. Neptune completes a full orbit around the Sun every 165 years; it has barely completed a full orbit since it was discovered. The planet completes a rotation around itself every 16 hours. The planet has a ring system (much smaller than Saturn's), 14 moons discovered until now, and a windy atmosphere rich in hydrogen, helium, and methane. Neptune, like Jupiter, has a spot visible on its surface; it is a blue spot believed to be a hole in the methane layer.

Triton is the largest of the Neptune Moons and seems to be the only large moon in the solar system that orbits its planet in the opposite direction of its rotation. Some explanations indicate that the moon might have been formed elsewhere and then it was attracted to the planet by its gravity. However, this is not the only theory out there. Neptune has been visited by Voyager 2 after a 3.5-year journey from Uranus. I note that the time of the trip is not a precise measurement of how far a planet is because of the many factors that should be considered, such as acceleration, the planet's position in its orbit, and many other variables.

Pluto: A Dwarf Planet

I would not usually have a section dedicated to Pluto since it is NOT a planet. However, given the historical and informative importance of the matter, I decided to dedicate a section to it.

Pluto is the only planet to be... Oops, it is not a planet. Ok, let me rephrase that. Pluto is the only trans-Neptunian object (beyond the orbit of Neptune), that was thought to be a planet. It was discovered by American Clyde William Tombaugh, in 1930. At the time, and until 2006, it was the 9th and most outer planet in the solar system, at 39 au from the sun, in the Kuiper belt. All planets have an elliptical orbit, but Pluto's orbit is even more oval-shaped than the others. Pluto's orbital period is 248 years, and since its discovery, it has not yet completed a full orbit around the Sun. The temperature on Pluto is around -230°C. Even though Pluto is 6 times smaller than Earth, it has five natural satellites (moons). Charon is the largest one and is almost half the size of Pluto.

In 2006, the International Astronomical Union decided to demote Pluto from a planet to a dwarf planet. To explain why Pluto was demoted, let me first clarify what was defined, in 2006, as a planet. Before 2006, there was no actual definition of a planet, and the name was traditionally assigned when a planet-like object was discovered. However, astronomers started finding objects in the Kuiper Belt that looked like planets. For example, Eris was discovered in 2005 and was going to be considered the 10th planet, and Sedna was found in 2003 but did not have a precise classification. In total, there are five known similar dwarf planets in the solar system. All these objects could not be classified as planets since they differed in size, distance, and orbit shape. Therefore, the definition of a planet was set.

Let's see if Pluto fits all the set criteria of a planet. A planet is a celestial object that is in orbit around the Sun. Pluto: check. A planet should have enough mass to assume a round shape. Pluto: check. Finally, a planet should have cleared the neighborhood around its orbit, and this is the main criterion that Pluto did not make. The path of Pluto in the solar system is filled with debris such as asteroids, planetesimals, and other objects. This is one of the main reasons why Pluto is no longer considered a planet. Other reasons emerged, such as Pluto's small size and the fact that it is a rocky object positioned after several gaseous planets. Only one mission was sent to Pluto, New Horizon, and it was the fastest human-made probe to be launched. It was sent in 2006 and reached Pluto in 2015. Before the photos of the probe, we only had fuzzy blob pictures of Pluto. New Horizon took the first clear pictures of Pluto and discovered a heart-shaped region of ice, as well as several mountains of iced water. New Horizon is now wandering in the Kuiper belt, collecting data on trans-Neptunian objects.

Kuiper Belt and Comets

The Kuiper belt is an outer region of the solar system made of trans-Neptunian objects surrounding the solar system. It is located at 30 au from the sun and extends to more than 50 au from the Sun. It is beyond the orbit of Neptune, and it contains asteroids, comets, dwarf planets, and other icy objects. The astronomer Kuiper predicted the existence of the Kuiper Belt, and the first objects in it were discovered five years later (other than Pluto, which was found before). Pluto is inside the Kuiper belt, and it is the largest dwarf planet discovered to date. Other dwarf planets also exist in the

Kuiper belt, like Eris. Eris is the second largest object in the Kuiper belt, but it is in fact the most massive.

In 2023, astronomers discovered a ring around the dwarf planet, Quaoar, which came as a surprise. When it comes to small and far objects, like asteroids and dwarf planets, direct observations are difficult as the objects studied are very small. An occultation method can be used instead. An occultation occurs when an object passes in front of a star, as observed from Earth. The shadow of the object can be used to study its morphology. In the case of Quaoar, astronomers spotted its main shadow. However, two smaller shadows, before and after the planet's passage were also seen, indicating the presence of the ring. One of the other objects discovered beyond the Kuiper belt is Sedna; a dwarf planet discovered in 2003. Sedna has a very elliptical orbit, ranging from 76 au to an astonishing 937 au from the Sun. This dwarf planet takes 11,400 years to complete a full orbit around the Sun. In comparison, it takes Sedna more time than the rise of civilization to complete an orbit around the Sun. Other particular objects that can be found in the Kuiper belt are comets, in particular, short-period comets.

First, what is a comet? Comets are celestial objects composed of a core of ice and rocks. They were formed by the clumping of gas and dust in the cold region of the system during their formation. They usually orbit the Sun in very elliptical trajectories. The Sun is one of the focal points of the ellipse. During their orbits, when comets come close to the Sun, their icy surface evaporates and leaves behind a tail of particles. When observing a comet, this tail is the cloudy shape we see in the sky. Every time a comet approaches the Sun, a part of it evaporates until it all eventually disappears.

The most famous comet is Halley's comet. It has an orbital period of 76 years and is believed to have originated from the Kuiper belt. Halley is an English astronomer who looked at previous reports of comets passing in 1531, 1607, and 1682. He realized that this was the same comet, with a ~76-year orbital period, and predicted its next approach near Earth to be in 1758. His predictions were accurate, and the comet approached Earth during that year, but Halley did not live to see it. With an orbital period of 75 years, a human can only hope to witness this comet once or twice before they die. Halley's comet visited Earth last in 1986, seven years before I was born, and it is expected to return in 2061. If I am lucky enough to be alive, I will witness it at 68 years old. Other comets from the Kuiper belt can take up to 200 years to complete an orbit.

Not all comets originate from the Kuiper belt, some come from far beyond, and some have an orbital period of 250,000 years, like Comet West. The latter passed through the inner solar system in 1976. Civilizations were probably not yet established on Earth during its previous passage. Humans sent several probes to perform close encounters with comets, and the last one was the Rosetta spacecraft (ESA) that visited Comet 67p in 2014. The lander, Philae, was planned to land on the surface of the comet to conduct the most detailed comet study to date. Unfortunately, Philae had a bumpy landing and was stuck in a dark crevasse where it ran out of power. However, Rosetta still produced valuable data on the comet. Deep Impact (NASA) was sent to fly by comets 9 P/Temple and 103 P/Hartley. In 2005, Deep Impact released an impactor that collided with the 9 P/Temple nucleus, resulting in excavated debris that was studied by the Deep Impact instruments. The results showed the comet to be dustier and less icy than expected. Sometimes, as Earth travels around the Sun, it enters a region where a comet has crossed before. This region is full of small particles that the comet has left behind. The particles enter the Earth's atmosphere but disintegrate due to friction with air. These are meteor showers that light up our skies. August is usually the peak for meteor showers, as it is when we cross the path of comet Swift-Tuttle, causing the Perseids meteor shower.

Planets 9, X and 10

Astronomers are observing the movements of trans-Neptunian objects and noticing dynamical perturbations in their motion, which could be explained by the existence of a 9th planet. This planet, known as planet 9, (sometimes called planet X as the naming rights belong to the discoverer) would be located somewhere between 400 and 600 au from the Sun (400 to 600 times the distance from Earth to the Sun) and is predicted to have 5 to 10 times the mass of Earth. All these predictions are made just by observing other objects.

Astronomers are looking for Planet 9, but have yet to be successful. Alternate hypotheses suggest that the planet does not exist and that the observed behavior of these trans-Neptunian objects is biased. Others suggest that additional dynamical perturbations can explain the movements of these objects in the solar system. We will have to wait and see what astronomers discover. It is undoubtedly an exciting time to be alive and witness these fantastic discoveries.

Planet 10 has a similar story. It is another hypothetical planet that may exist beyond Pluto's orbit, with a mass ranging between that of Mars and Earth, much smaller than Planet 9.

The Heliosphere

The heliosphere is the outermost part of the Sun's atmosphere. The Sun's atmosphere extends well beyond the orbit of Neptune. The Sun's heliosphere is also its magnetosphere, a magnetic field bubble that is created by the solar wind, a protective barrier shielding the planets and the solar system from interstellar radiation. The heliosphere extends up to 100 au away from the Sun.

The Oort Cloud

The Oort cloud is a theoretical concept. It is supposedly a region surrounding the entire solar system, but it is very far away from the most distant object in the system. The Oort cloud was first conceptualized by Dutch astronomer, Jan Oort, in 1950. The cloud is the house of icy planetesimals and bodies, like comets with long orbital periods. It is a region that starts at approximately 2,000 au from the Sun and extends to 50,000 au or 100,000 au, nearly a third of the distance to the nearest star. The Oort cloud is the most outer region of the solar system, and it is heavily affected by interstellar wind and gravitational pulls from other stars. Until today, there is no observational evidence of the Oort cloud, but this might change in the future.

The Future of the Solar System

As mentioned at the beginning of this chapter, the solar system was formed 5 billion years ago from the gravitational collapse of a cloud of gas and dust. In the beginning, it was very chaotic. The Sun was a protostar, with a concentration of matter in the center surrounded by a pancake of accumulated matter; an enormous infernal tornado of matter falling into the proto-Sun. Inside this tornado, the collisions started forming planetesimals and planets. The rocky planets formed closer to the Sun, where it was hot, while the gaseous planets formed farther away where the temperature was low enough to allow condensation. Several theories explain the existence of rocky planets on the inside, and gaseous planets on the outside. From which, the migration theory explains that larger planets formed close to the Sun and migrated outwards. But let's not complicate things and simply stick to the clear parts of the story. Earth was a red-hot magmatic planet that was bombarded by countless asteroids and planetesimals. Asteroids and planetesimal bombardments were common at this early chaotic stage of the solar system's formation. Planets eventually cleared their orbit from these objects, and things started to cool down about half a billion years, to one billion years later. In the meantime, solar winds blowing from the Sun were also cleaning up the solar system, sweeping most of the gas and dust outwards beyond the heliosphere. Things became more stable, and today, the solar system is considered a more or less stable middle-aged system. Of course, planets still encounter asteroids and comets from time to time.

The future of the solar system is determined by the Sun, which will eventually reach the end of its life around 5 billion years from now. By then, the Sun will start swelling until it possibly reaches the orbit of Earth, engulfing the inner rocky planets. Earth will be incarcerated by the Sun's heat. The Sun will continue to swell until its outer envelopes fade into clouds of gas and dust and it cools down. Its core will remain hot and dense, in the form of a white dwarf, approximately the size of Earth. I will give more information on white dwarfs in the next chapter.

Chapter 4

Stars

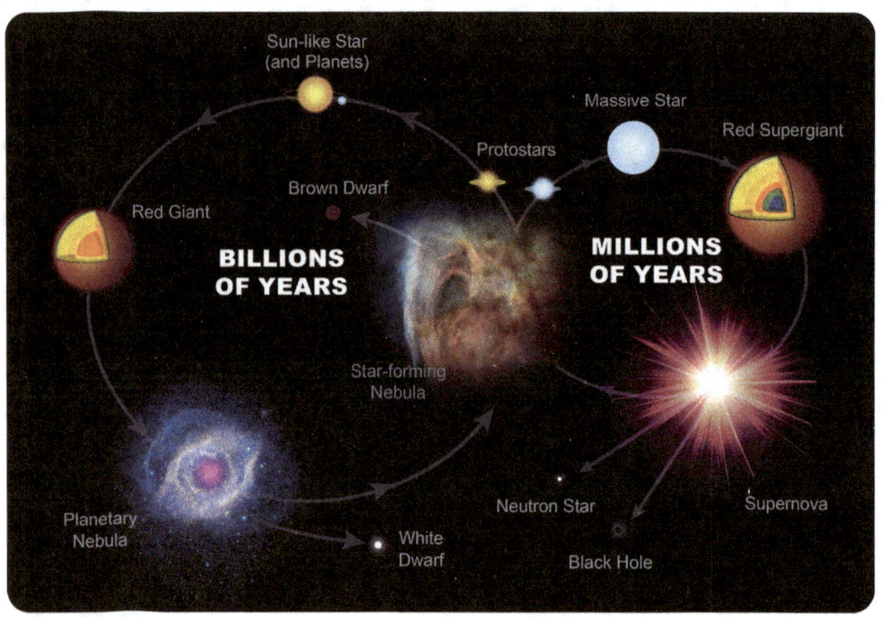

Sun-like Star
(and Planets)
Massive Star
Red Supergiant
Protostars
Brown Dwarf
Red Giant
BILLIONS
OF YEARS
MILLIONS
OF YEARS
Star-forming
Nebula
Planetary
Nebula
White
Dwarf
Neutron Star
Supernova
Black Hole

In the previous chapter, we discussed the solar system, our home, centered around the Sun, a star. The Universe is much bigger than the solar system; in comparison, the solar system is just a small dot in the galaxy. A galaxy is a large group of stars, gas, and dust. Our galaxy, the Milky Way, is estimated to host between 100 to 400 billion stars. At least half of these stars are thought to have an entire system of planets around them. There are trillions of galaxies out there. In this chapter, we will talk about stars in general. I want to remind you that the Sun is a star, just like the ones we will discuss in this chapter. It is also imperative to know that each star system is different, and the properties of stars differ in type. Proxima Centauri is the closest star to the Sun, which is 4.2 ly away and is the closest star to us, after the Sun. Other stars in the galaxy can be as far as thousands of light-years away. The Universe without stars will only have hydrogen and a little helium. The stars produce the remaining elements like Carbon, Oxygen, metals, and everything else. These elements are also the ones that form life. We are stardust. You might have heard this before. It is an inspiring statement, but it is more than that. It is a fact. Stars are also the progenitors of one of the most mysterious objects in the Universe: black holes. So, what is a star? What are the different types of stars? What is the life cycle of a star? What is their composition? Let's find out.

: ●

Stars in General

Stars twinkle at night. Stars are huge, but they are also very far, and the light they emit appears to be coming from a single point at an infinite distance. Starlight encounters several layers in the atmosphere with different densities and temperatures that refract in different directions. The light will sometimes travel toward our eyes and sometimes refracts in other directions. The amount of light entering our eyes continuously fluctuates from bright to faint, and due to this fluctuation, the light appears to twinkle. This is your answer to why the stars twinkle. This effect is not visible when observing a planet. The light from a planet comes from a much closer object that appears larger; it is similar to a large number of point-like sources. Some light from these sources refracts, but the light entering our eyes does not fluctuate, or it fluctuates only a little, as there will always be light from multiple point-like sources that enter our eyes. Without using any telescope or positional information, we can identify if a celestial object is a star or a planet, with our naked eye, simply by watching if it twinkles or not.

The first thing to remember is that, like the Sun, all stars are spheres of gas. The spherical shape, like the shape of planets, comes from the gravity dynamics that are pointed toward the center. Everything in the Universe that has a mass has gravity; from the biggest star to the smallest particle of gas. When there is a massive cloud of gas in the Universe, hydrogen gas in particular, the gas particles will attract each other and start forming a sphere. Gravity will become stronger when there are more gas particles, and when gravity becomes stronger, the particles will be attracted to each other further until they collide and create friction. When the frequency of collisions increases, friction will also increase, and the temperature will rise with it. This rise in temperature will lead to more movement, and thus more friction of particles. This in turn will increase the temperature to millions of degrees. When such temperatures are reached, fusion is ignited. Two atoms of hydrogen fuse to give one atom of helium, twice as heavy as hydrogen. Except that the atom of helium is slightly less massive than the total mass of the initial hydrogen atoms. The rest of the mass is converted into energy, particularly radiation. This is how stars give us light and heat; through the radiation emitted from nuclear fusion.

Near the end of its life, a star will have sufficient energy to fuse helium into more massive atoms. Atoms of helium fuse resulting in heavier elements like carbon, and carbon, along with other elements, then fuse to give oxygen and other elements, and so on. The more energy (mass) a star has, the heavier the atoms it can fuse at the end of its life. All of this happens in the heart of the star; the heart that is surrounded by other layers that we will soon explore.

Remember that while a star looks like a ball of flames, it is not combusting its fuel, but rather fusing it; providing light and warmth through nuclear reactions. So, a star burns its fuel mainly through nuclear reactions. The star's core is where the action (fusion) happens. A core can have several layers depending on the age of the star. The inner layer (here, we are talking about the layer of the core and not the layers of the star) contains the heaviest fused material. The surrounding layers contain progressively lighter materials. In the outer layer of the core, hydrogen (the lightest element) fuses into helium (the second lightest element).

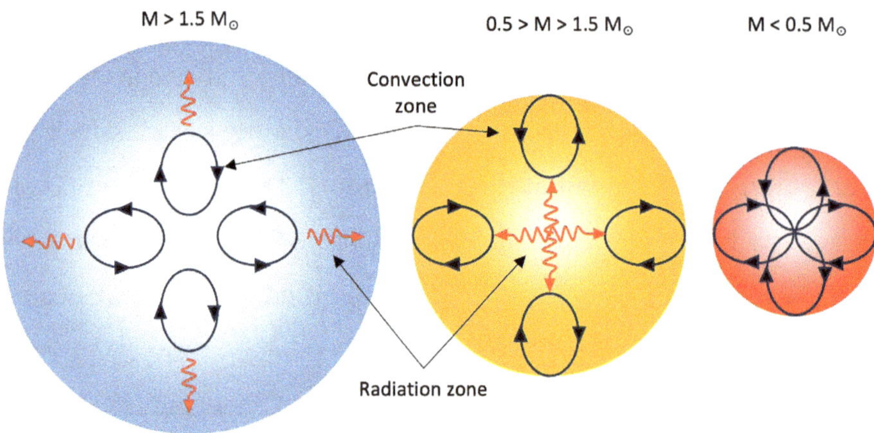

Figure 4.1

As shown in Figure 4.1, around the center of Sun-like stars, there is a massive layer of gas where the energy is transferred from the core to the upper surface, through radiation, called the radiation zone. After the radiation layer, an additional layer called the convection zone, transfers the energy through convection; the hot gas goes in circles, moving heat from the inner zone to the outer zone. In more massive stars, the convection zone is around the center, while the radiation zone is outside. In smaller stars, the energy transfer is only done through convection. On the edge of the convection layer is the photosphere; the surface of the star that we see. The light and heat from the star are radiated from the photosphere. A star's atmosphere is usually composed of two layers: the chromosphere, which extends a few thousand kilometers, and the corona, which extends over several million kilometers. The corona contains plasma composed of charged particles and can reach higher than one million degrees in temperature.

$$\vdots$$

Classification

Stars are very distant and relatively small when compared to galaxies, for example. Consequently, until only a few years ago, we could never capture a photo of any star's photosphere (surface) except the Sun's. Some of the most sophisticated astronomical techniques (called interferometry, discussed in Chapter 11) are applied to some of the most sophisticated telescopes in the world to get blurry images of the biggest and closest stars to us.

We usually study stars by looking at their composition and the amount of light they produce. For example, the color of a star is an indicator of its temperature. Cooler stars have colors closer to red, while hotter stars, which are also bright, have colors closer to blue. There are seven types of stars: O, B, A, F, G, K, and M; which are classified according to their surface temperature and spectral features. Without going into many details, note that O-stars are blue and the hottest stars in temperature, and the classification goes down until we reach the M-stars, which are red and the coolest in temperature. This is the Morgan–Keenan (MK) classification system that, in fact, contains many more sub-types and expansions. For example, each letter is subdivided into numerical digits ranging from 0 to 9, with 0 being the hottest and 9 the coolest. It is essential to note that M-stars are the most common in the Universe. Nearly 75% of the stars in our neighborhood are M-stars. In fact, most stars are red dwarfs of the M-type, with less than half the mass of the sun. The Sun is a G2 star with a surface temperature of 5,500°C (5,800 Kelvin (K)). It is estimated that around 5% of the stars in the galaxy are more massive than the Sun, and some stars can reach more than 100 times the mass of the Sun.

Lifecycle of Stars

When a star is formed, it is called a protostar. It resembles a pancake with a luminous core and a disk of accreted material. Now that I think of it, a better analogy would be a fried, sunny side up, egg. In the center, the star is born, and most of the material is accreted. The remains then form the surrounding planets and small objects. After the protostar stage, the central life of a star starts; it is called the main sequence. Most of the stars we see are in the main sequence, which is the longest stage of a star's life. A lot of a star's properties, such as its lifetime and lifecycle, are dependent on its mass. The mass of the Sun is 2*10^30 kg, which is around 333,000 times the mass of the Earth. Since it is not easy to deal with these numbers, we will use the mass of the Sun as a reference. A star with an equivalent mass to the Sun will live for 10 billion years, or to be more precise, it will remain as a main sequence star for 10 billion years. A star with 10 times more mass (10 solar masses) will only live for 30 million years. As the star gets heavier, its lifespan is significantly reduced. A star that weighs 100 solar masses has a lifespan of a 100,000 years. In contrast, the less the star's mass is, the longer it lives. A star that is 0.1 solar mass can live for several trillion years. Even though a massive star has more fuel, its lifespan is smaller because it burns much faster.

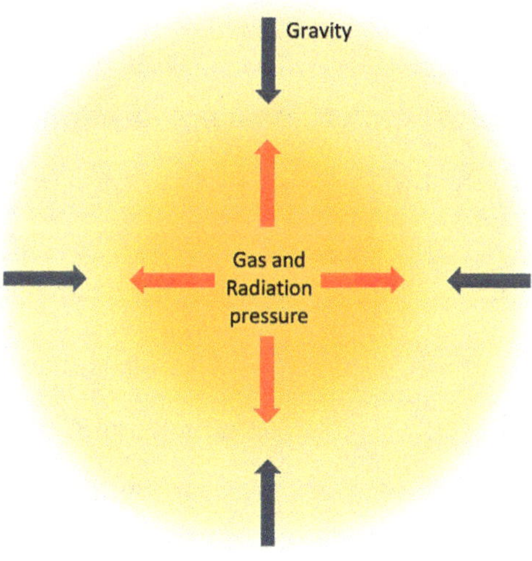

Figure 4.2

It is essential to know that a star reaches equilibrium due to two equating forces: The force of gravity that pushes everything inwards, toward the center, and the force (pressure) of the heated gas that pushes outwards from the center. The latter is also assisted by an additional force generated from the pressure of the energy released by the nuclear reactions in the core and carried outwards. So, there is an inward and an outward force, and these two forces equate, keeping the star's shape spherical. A representation is provided in Figure 4.2. When the thermodynamics of the star change, the gas pressure changes, and the star loses this equilibrium. This is the beginning of the end of the life of a star; the end of the main sequence stage. As the hydrogen is converted into helium, the core becomes more and more massive. The increasing gravity contracts the core, which heats further and further. At the end of the life of a star, when all the hydrogen is depleted from the core, the core reaches temperatures high enough to fuse heavier elements like helium. Helium fusion then begins. The star's temperature increases further, breaking the equilibrium mentioned earlier. The star then starts to swell becoming a red giant star.

Other stars can become other types of giants. Massive stars will even become supergiant stars. In fact, a classification using Roman numerals is used in the Morgan-Keenan scheme to designate, what is called, its luminosity class. Class I is supergiants, II is bright giants, III is regular giants, IV is subgiants, and V is main-sequence stars, in addition to other classes that I will not mention here. The Sun is a G2V star designating a main sequence star with a surface temperature of 5,500°C (5,800 K). If we place a red supergiant giant in place of the Sun, its circumference can reach beyond the orbit of Mars. When the nuclear fusion fuel is depleted, the nuclear reaction stops, and the force of gravity takes over. Gravity shrinks the core until it becomes a dense ball of material called a white dwarf. Not all stars result in white dwarfs. Lighter stars, with less initial mass, tend to become white dwarfs. On the other hand, heavier stars undergo different transformations: some give neutron stars, and others give black holes. The outcome of a star is linked to its initial mass, determining whether it ends as a white dwarf, neutron star, or black hole. These objects are called compact objects in the astrophysical community since they are very compact. I like to call white dwarfs, neutron stars, and black holes, dead stars, although some might not agree with this naming since these objects are far from being dead or inactive.

In this section, I have provided the general stages of a star's life. In the next section, I will describe the life cycle of stars by mass category and provide an overview of the three types of dead stars. The evolution of stars and their classification is summarized in a diagram that astrophysicists often refer to as the Hertzsprung-Russell diagram illustrated in Figure 4.3.

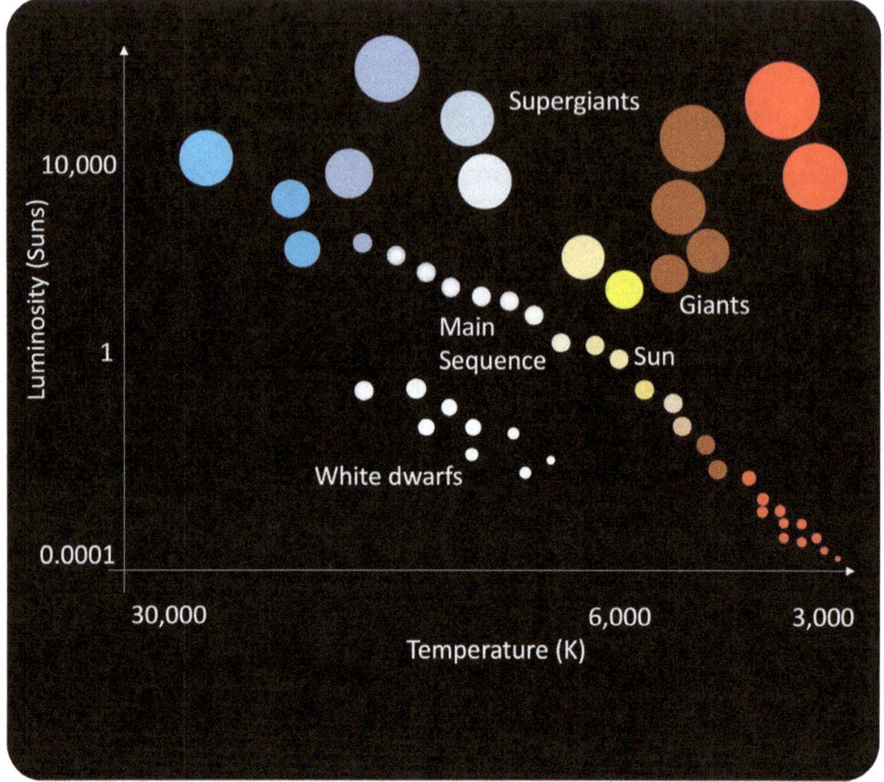

Figure 4.3

Lifecycle of Stars Versus Mass

Mass < 0.08 Suns

Let me start this section by talking about the stars that don't have enough material to become stars; they are called brown dwarfs. As explained before, the force of gravity is responsible for forming a star. It is responsible for bringing the gas atoms together, creating friction, and raising the temperature to ignite nuclear fusion. Sometimes, there is not enough material; hence not enough gravity to raise the temperature to the millions of degrees needed for fusion to start. In that case, the star will never ignite and it will remain a ball of gas and dust. It is a failed star, an object with a mass between a giant planet and a small star. Usually, this happens to stars with masses less than (<) 0.08 the mass of the Sun, or 80 times the mass of Jupiter. The lowest limit of the mass of a brown dwarf is more than (>) 10 times the mass of Jupiter.

0.08 Suns < Mass < 8 Suns

The minimum mass for a star to become a star is 0.08 solar masses. Stars between 0.08 and 8 solar masses have a similar life cycle. However, their lifespan differs, with the lighter ones living much longer. These stars start as protostars before entering the main sequence. In their core, hydrogen is fused into helium with a core temperature of around 15 million °C. Hydrogen fusion starts in the shells surrounding the helium core when the hydrogen is depleted there. Inside the core, helium requires much higher temperatures to fuse, in the order of hundreds of millions of degrees. These temperatures are reached due to the contraction of the core, and helium starts converting into carbon at around 100 million °C. The outer layers of the star start expanding due to the increasing heat. This is the beginning of the death of the star where it escapes the main sequence and becomes a red giant. This phase lasts for a few million years. Inside the core, Helium and Carbon interact and give oxygen, but at a certain point, the core runs out of helium.

For small-mass stars described here, the gravitational energy is insufficient to raise the temperature enough to fuse atoms larger than oxygen. Consequently, the fusion stops in the core, leading to the contraction of the core. Simultaneously, the star's outer layers expand until they become light shells of gas and dust, known as planetary nebulae. The core then becomes a white dwarf.

White Dwarfs

Inside the star's core, once the fusion process stops, its energy opposing gravity is no longer present, allowing the gravitational force to become dominant. The core continues to shrink, causing the atoms to come closer and closer together. When the atoms inside the core become too close, the electrons of these atoms are forced to repel each other. The electron repulsion force is now the force opposing the gravitational collapse. A new equilibrium is reached, with gravity pushing toward the center, and the electron repulsion force pushing outwards. White dwarfs are luminous and hot (100,000°C) due to residual heat, but fusion does not take place inside a white dwarf. A white dwarf is very dense since the atoms are very close. It contains more than half of the star's mass and is the size of Earth. Imagine 50% of the mass of the Sun (167,000 times the mass of Earth) compressed in a ball the size of Earth; This is considered dense. One tablespoon of white dwarf material would weigh as much as an adult elephant.

We do not know what a white dwarf will later become and how it might evolve. A white dwarf's lifespan can extend to more than 10 billion years, and it is believed that when a white dwarf no longer emits energy, it becomes a black dwarf. We have never seen a black dwarf before, but this is understandable since they do not supposedly emit much light or heat. Moreover, if the theory is true, then it will take a long time for a white dwarf to become a black dwarf, a time comparable to the age of the Universe. So, black dwarfs might not have even formed yet.

Mass > 8 Suns

Massive stars, with more than 8 solar masses, have a life similar to the lighter ones, but with a very different death. The stages are similar until the star reaches the end of the main sequence. First, it is a protostar, then a main sequence star with hydrogen fusion in the core. However, when the hydrogen in the heart of a massive star is depleted, the star's death is different. Helium starts fusing in the core, and hydrogen in the surrounding layers. The star becomes a red supergiant. Larger atoms, like carbon, undergo nuclear fusion and give oxygen. Massive stars have more energy to fuse more massive atoms. Therefore, the fusion does not stop at oxygen when the helium is depleted in the core after a few million years. More and more massive elements fuse to form even more massive elements, such as oxygen, neon, silicon, sulfur, and iron. This is where the fusion stops, at iron. Iron is a very stable element, and fusing iron to get heavier elements requires enormous amounts of energy that even a massive star cannot provide. So how are heavy materials like Gold, Silver, and Platinum produced? You will find out in a second.

Iron is called star killer because when a star starts forming iron in its core, it reaches its final stages —its death bed. So now, the star has an Iron core surrounded by layers of progressively lighter material. Around the Iron core, there is a layer of silicon fusing. Around this layer is a layer of magnesium fusing, and it goes on until it reaches the outer layers of the core, where there is a layer of carbon fusing, surrounded by a layer of helium fusing, surrounded by the final layer of hydrogen fusing. The iron inside the core cannot fuse, so the fusion stops. This will cause the Iron core to implode very quickly under the effect of gravity. The implosion of the Iron core destabilizes the entire star. Material falls, in a split second, toward the center with speeds reaching a quarter of the speed of light. Material coming from all directions bounces off, producing an extremely violent shockwave, causing the star to explode; this is called a Supernova. The explosion raises the temperature of the star drastically providing the energy needed to form heavy metals, like Gold and Platinum. The remains of the star, gas, and dust shatter in the Universe due to the explosion. Supernovae are one of the most cataclysmic cosmic events. The energy released during a supernova is 100 times greater than what the Sun will radiate during its entire lifespan.

We will be talking about supernovae again in Chapter 9. After a star undergoes a supernova, if the core of a star survives, it will become either a neutron star or a black hole.

Neutron Stars

Neutron stars will form when the mass of the star's remaining core is between 1.2-1.4 and 2-3 solar masses. Under the effects of gravity, the imploding core shrinks until the atoms become too close. This time, the force of gravity is much greater than that of white dwarfs. The electron repulsion force is not enough to hold the stability of the imploding core, and it is overcome by the force of gravity. The core continues to shrink, and the atoms get closer and closer together. The nuclear repulsive forces between the neutrons in the star become too strong and equalize the force of gravity, providing equilibrium to the neutron star.

A neutron star is characterized by its substantial density and small size. A neutron star heavier than the Sun is the size of a city; imagine just how dense this is. A neutron star is denser than a white dwarf. One teaspoon of neutron star material would weigh as much as a mountain. A neutron star has a powerful magnetic field that can alter the structure of the atoms on its surface. Regular neutron stars have a magnetic field that is a trillion times stronger than the magnetic field of Earth. When a neutron star has a super strong magnetic field, around 1,000 times stronger than the average neutron star, it is called a magnetar. Neutron stars and magnetars spin fast; some can spin several hundred times in less than a second. Imagine a city spinning on its axis several times in one second. The velocities and magnetic fields reached by these celestial objects put them at the top list of the most extreme and violent objects in the Universe. A neutron star emits jets with strong radio emissions from its poles. When the neutron star is spinning rapidly, and the collimated emission crosses Earth, it is seen as a flashing object in the sky called a pulsar.

Black Holes

In the case where the mass of the dying star's core is greater than 2 to 3 solar masses, a black hole will form. Whether it is the electron repulsion force or the neutron nuclear forces, no force can stop the implosion of the core due to its gravity. The core will continue to collapse until it becomes a singularity. At a specific point, gravity becomes so strong that even light cannot escape. All the light from a particular region around the singularity cannot escape, and any object inside this zone will be attracted toward the center. The edge of this zone is called the event horizon. It is represented in Figure 4.4. A black hole is formed. Inside the event horizon, no object or particle can escape the gravity of the black hole, and nothing can come back once it crosses the event horizon, not even light (hence the word black). Roger Penrose won the 2020 Nobel Prize in physics for proving that black hole formation is a robust prediction of general relativity.

To escape the gravity of any object, such as Earth, a certain amount of energy must be exerted in the form of speed. Rockets need to move fast to escape Earth's gravity. The closer we are to Earth, the greater the speed required to escape its gravity. The greater the gravity, the faster we should move to escape the object. A speed of 11.2 km/s is required to escape the surface of Earth. A speed of 60 km/s is required to escape the surface of Jupiter. A speed of 615 km/s is required to escape the surface of the Sun. A speed of 300 000 km/s (the speed of light) is required to escape at the edge of a black hole event horizon, but at a closer distance to the black hole, even the fastest thing in the Universe, light, cannot escape the gravity of the black hole anymore. In other words, we need to move faster than light in order to escape the black hole, but nothing is faster than light.

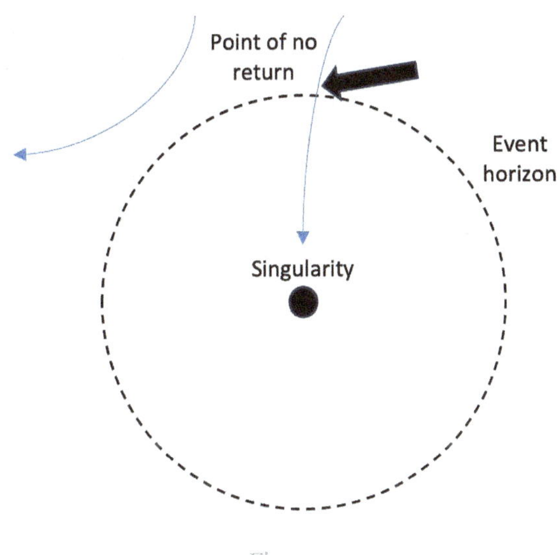

Figure 4.4

The distance from the center of the black hole to the event horizon is called the Schwarzschild radius of the sphere. It depicts the event horizon of the black hole. In another way, you can imagine that the black hole creates a curve so deep in space-time that nothing can escape it. This is why black holes are black; seeing them is impossible because they have no electromagnetic emissions. Sometimes the surrounding material around them emits light and we can visualize the black hole by looking at the shadow it creates. Keep in mind that we are talking about stellar black holes, which are created by the core collapse of giant stars. The mass of Stellar black holes ranges between three and several dozen solar masses. This is important to know since there are other types of black holes. We have supermassive black holes found in the center of galaxies; they are a hundred thousand, to several billion solar masses, but more on that later. We have intermediate-mass black holes that range between a hundred to several thousands of solar masses. Intermediate black holes have only been detected very recently, but more on that in Chapter 9. We also have primordial black holes (with a wide mass range) that formed directly after the Big Bang, but they are still hypothetical at this point. Finally, there are hypothetical microscopic black holes that can also be traced to primordial origins.

Today we do not have the physics to explain what happens inside a black hole, however, we know what happens around it. For example, as described in the previous chapters, when gravity is strong, time slows down. If you are falling into a black hole and your friend is watching, from their perspective, your time is slowing down until it stops when you reach the event horizon. This is why your friend will never see you crossing the event horizon since it will take a very, very, very long amount of time to travel the last few inches. However, from your point of view, nothing has changed. In a finite amount of time, you will eventually reach your destination, and probably your demise.

That's right. If you attempt to cross the event horizon of a stellar black hole, you will probably die.

Let's consider a plastic mannequin falling into a black hole. The gravitational tides from the black hole are so strong that if the mannequin goes in feet first, its feet will get pulled much stronger than its head (because they are closer to the black hole). Eventually, the mannequin's body snaps, then each half splits in half, and so on. As if this is not enough, in addition to stretching, the mannequin's body will also undergo compression in the perpendicular direction of the fall. Basically, the mannequin will stretch and disintegrate to the level of atoms while falling into the black hole. This process is called spaghettification. Don't worry. The closest discovered black hole to Earth is 1,500 ly away, and it will take humanity millions of years to reach it with our current technology. Black holes do not emit electromagnetic radiation and are invisible to the observer. If we want to observe a black hole, we usually target the surrounding matter to image it. Although the existence of black holes was established a while ago, the first images of a black hole did not appear until 2019, when scientists combined data from radio telescopes all over the globe, simulating an Earth-sized telescope.

Steven Hawking, a renowned physicist, theorized in 1974 that a black hole could emit radiation through, what is called, Hawking radiation. In quantum physics, a pair of matter and anti-matter particles may be created simultaneously without priors. Quantum physics is weird!

The created particles will then annihilate each other. In a situation where the particles are created near a black hole, and one of the particles is pulled inside the event horizon while the other is outside, the second one escapes in the form of radiation: Hawking radiation. So, black holes can radiate, and as Hawking says, they are not so black. Hawking radiation process causes the black hole to lose energy slowly and lose its mass. As it loses mass, it radiates more and more and eventually disappears after a significant amount of time. Therefore, through Hawking radiation, black holes can evaporate. This theory has many paradoxes that still need to be solved. To date, Hawking radiation has never been detected by astronomers.

How to Study Stars

Stars, and most celestial objects, are very far away and out of reach.So how have we learned all this information about stars?

Nowadays, there are several observational methods used to study stars. The most common ones are spectroscopy, photometry, and astrometry. Spectroscopy studies the composition of the light emitted by a celestial object. A spectrum is a measurement of the amount of light in function of the energy (or frequency) of the incoming photons (more info in Chapter 11). Electromagnetic radiation interacts with matter; sometimes, matter absorbs parts of the light spectrum, and other times it will add to it. By studying the composition of the light and looking at the spectrum, we can conclude what type of elements interacted with the light as shown in Figure 4.5. This is how we know the composition of a star, by identifying the materials it contains that interact with the light we receive.

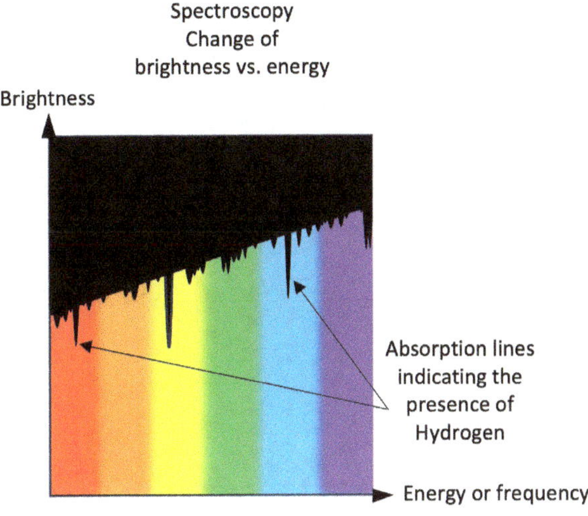

Figure 4.5

Photometry is the science of measuring the amount of light from an emitted source and compute light curves (more info in Chapter 11). A light curve is the measurement of the amount of light emitted by a celestial object in a certain energy band in function of time as shown in Figure 4.6. Photometry can be instrumental in measuring the properties of stars, like their distance, mass, temperature, and the temporal properties of their systems. Nowadays, photometry has become so advanced that it can even detect starquakes. Starquakes are seismic events that occur on the surface of a star, similar to earthquakes on Earth. These events can cause variations in the amount of light emitted from the star as it quakes. We can study its seismic activity by measuring how the light of a star dims and brightens. This science is called asteroseismology.

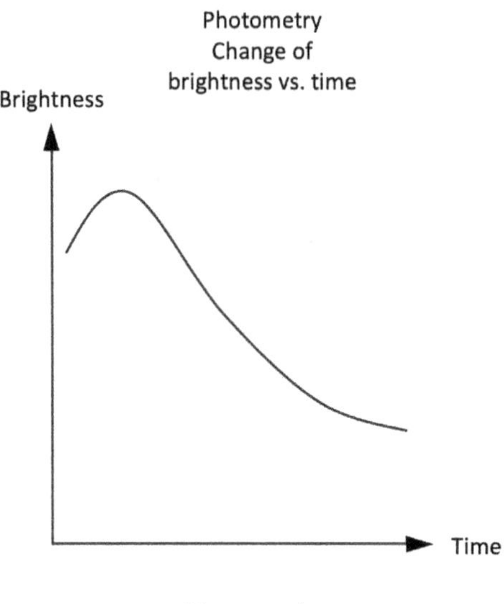

Figure 4.6

Lastly, astrometry is the study of the motion of celestial objects in the plane of the sky. For example, by taking several pictures separated by a certain amount of time, one can track how the position of celestial objects, like stars, varies with time as shown in Figure 4.7

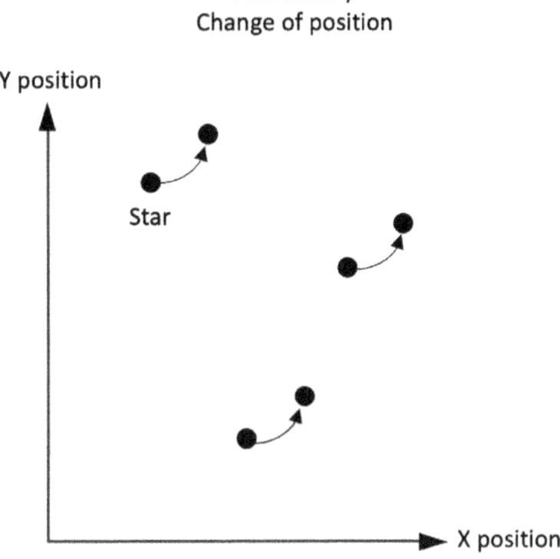

Figure 4.7

Stars in the Galaxy

Due to its close distance, the Sun is humanity's most known and studied star. However, what we know about the Sun is insufficient to understand all stars. For example, the Sun is a single star, with no companion stars. However, around half of the stars in the galaxy are binaries, meaning they have other companion stars. Some of them are in groups composed of three or more stars.

Stars are everywhere, whether they are single, accompanied, young, old, light, massive, dim, bright, alive, or dead. We can find large groups of stars in some parts of the galaxy; these groups are called stellar clusters. There are two types of stellar clusters: globular clusters and open clusters.

Globular clusters are a group of stars bound together by gravity. In general, they are spherically shaped and have a large concentration of stars in the middle. These stars can stay bound together for a very long time; therefore, globular clusters have a lifespan of several billion years. Globular clusters usually contain hundreds of thousands, to millions of stars. To date a cluster, we typically look at the age of the oldest stars.

Open clusters, on the other hand, are a group of stars that are less bound together. Because the gravitational bound is not strong in these clusters, they usually only survive up to half a billion years. They will relatively be quickly torn apart. Stars in an open cluster are young, and sometimes they might still be surrounded by the nebula (gas and dust cloud) that they were born in. Open clusters contain tens to hundreds of stars. You can easily find an open cluster in the sky, for example, the Pleiades and the Hyades, near the Orion constellation.

We are Stardust

After its birth, the Universe only contained hydrogen and some helium. Today, we can find more than 118 elements on Earth, like hydrogen, helium, carbon, nitrogen, oxygen, iron, gold, uranium, and all kinds of heavy metals. Some are very common, like hydrogen, while others are rare, like uranium. So how did we go from hydrogen and helium to 118 known elements? How did the Universe manage to create all these elements?

It did so by using its factories: the stars. As you have so far read in this chapter, the stars are responsible for creating elements heavier than hydrogen by fusing atoms in their cores; what cannot be forged in the hearts of stars is made at the end of their lifetime in supernovae explosions. So, the stars are responsible for the creation of elements beyond Hydrogen and Helium. The stars are also responsible for spreading these elements in the Universe. This is done in the final stages of their lives, either through a non-violent process of fading into a planetary nebula, or through extremely violent supernovae explosions. This, alongside the dynamic influence from galaxies, helps disperse gas and dust created by stars in the Universe. Gas and dust mix and form clouds.

As we saw in this chapter and the previous one, stars and planets are born in these clouds of gas and dust. The first generation of stars, the ones formed after the Big Bang, contained only Hydrogen and Helium (the only elements present at the time). During their lives, they created heavier elements and spread them in the Universe during their deaths. The clouds that formed afterward from the remaining material contained not only hydrogen and helium, but also a few heavy elements this time. The second generation of stars is formed from these new clouds of gas and dust; therefore, the second generation contains heavier elements than the first generation.

This cycle repeats itself, and the younger the star generation is, the more it contains heavy elements when they form. Although heavy elements are not only metals, the quantity used in astrophysics to determine how much a star has heavier elements than hydrogen is called metallicity. A star that contains many heavy elements has a higher metallicity number. The Sun is a third-generation star that contains many metals and heavy materials and has a high metallicity. This is because the initial cloud that formed the Sun contained these materials.

In the previous chapter, we learned that the planets are also formed by the same cloud that formed the Sun. Giant planets, small objects, and rocky planets, like Earth, were also formed by the elements in the cloud, particularly the heavy ones. This is why we have other elements besides hydrogen and helium on Earth. The raw materials were always there, since the beginning, and stars then manufactured the new materials.

Some heavy elements like carbon, hydrogen, nitrogen, oxygen, sulfur, and phosphorus are essential to life. In fact, these six elements are the main components that make up living things, and they are some of the most common in our bodies. They are also among the most common atoms in the Universe. So, we are made from elements produced by stars. To sum it up: We are stardust.

Chapter 5

Exoplanets

In this chapter, I discuss exoplanets. Exoplanets are planets that orbit stars outside our solar system. The Sun is a star with 8 planets orbiting it. Nearly half of the stars in the galaxy also have entire systems of planets orbiting around them. This may not be very instinctive, because until 30 years ago, we had never observed an exoplanet. For centuries we had thought that planets were only exclusive to the solar system. In summary, exoplanets exist. There are wonderful and mysterious worlds out there waiting to be explored. So, let's dive into the world of exoplanets.

How Exoplanets Are Discovered

Exoplanets are planets that orbit other stars. Stars are very far away, and today we can barely take a picture of the surface of a star. Knowing that planets are much smaller than stars, how do we discover exoplanets?

Direct and indirect methods are used to discover exoplanets.

Indirect methods look at how the existence of a potential exoplanet can affect its host star, and then a detection is made by looking at these effects. There are two widely used techniques to detect exoplanets indirectly.

The first one is the radial velocity method. In our solar system, the planets orbit the Sun because they are affected by its colossal mass, but the Sun is also affected by the mass of the planets. The same thing can happen in other star systems. When a giant planet orbits a star, the star will slightly wobble (move from side to side). For instance, when a planet is large enough, like Jupiter, its gravity will exert a considerable pull on the Sun. This effect causes the Sun to slightly wobble. Today we have methods to detect this wobbling. When a star wobble is detected, it could mean that a massive object is exerting a force on the star, generally a giant planet. The wobble effect can be detected by precise astrometry, the measurement of the position of the star in the plane of the sky. The wobble effect can also be detected in the perpendicular plane of the sky by measuring the radial velocity of the star, or in other words, how it moves along the line of sight of an observer on Earth. When something moves toward us, its color shifts to the blue, when it moves away, its color shifts to the red. An illustrative example of the two methods is given in Figure 5.1. To date, this method has allowed the detection of hundreds of exoplanets.

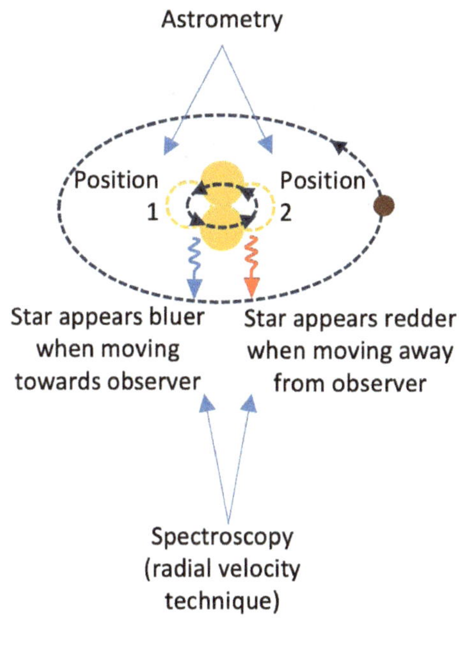

Figure 5.1

The second indirect method is transit photometry (illustrated in Figure 5.2), and it is responsible for thousands of exoplanet detections. Stars are enormously brighter than planets, and the light seen from a planet is mainly reflected from the star. However, when a planet orbiting a star passes in front of this star, from our point of view on Earth, the starlight dims slightly. By monitoring the amount of light received from a particular star, we can detect the passage of an exoplanet. Of course, the dimming of the light is so small that we need advanced instruments to see them. It would be comparable to detecting the dimming of a candle in a house in New York City from California. The bigger the planet, the dimmer the light will become. Transit photometry can also be accompanied by transit spectroscopy to further study the composition of the light of a star that travels through the atmosphere of a certain planet that passes between this planet and Earth. Transit spectroscopy provides a way to study the composition of the atmosphere of an exoplanet. It is useful to detect water vapor and oxygen, if any, in the planet's atmosphere.

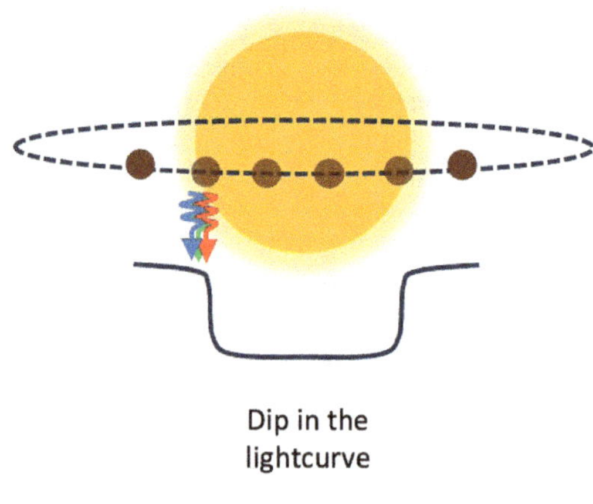

Figure 5.2

Another method to detect exoplanets is the microlensing method illustrated in Figure 5.3. When a massive star passes in front of another farther star, from our point of view, the star's gravity bends space-time, and the light coming from the faraway star gets bent. The closest star acts like a lens. If there is a planet orbiting the nearest star, the gravity of the planet will also bend space-time (with a smaller effect) and act like a microlens to the star behind it. When this effect is detected, it could be an indication of the existence of an exoplanet. This method is best for detecting planets far away from their star and is effective in detecting free-floating planets (planets that do not orbit a star, or that have been ejected from their star systems).

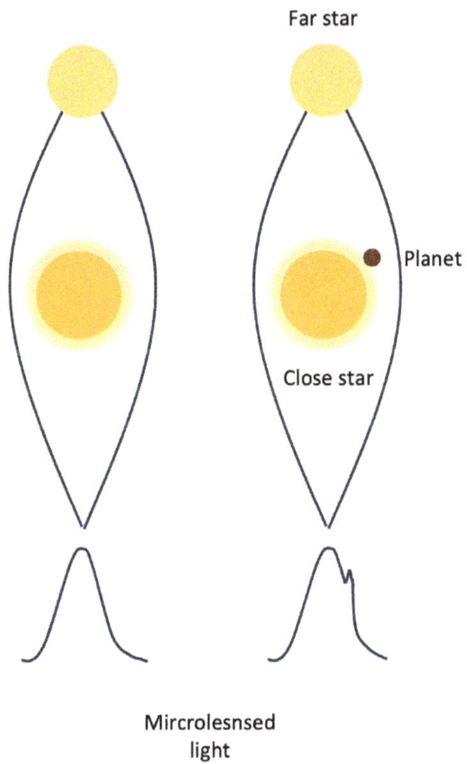

Figure 5.3

Direct methods, on the other hand, consist of taking an image of the star system and observing the planets. However, stars are so bright that their glare hides the planets. To be able to take a picture of these worlds, astronomers mask the star to stop the glare using a coronograph as shown in Figure 5.4. This is similar to what happens during a solar eclipse when the corona of the Sun becomes visible to us when the Moon stops the glow of the Sun. The imagery method works best on large planets that are far away from their star.

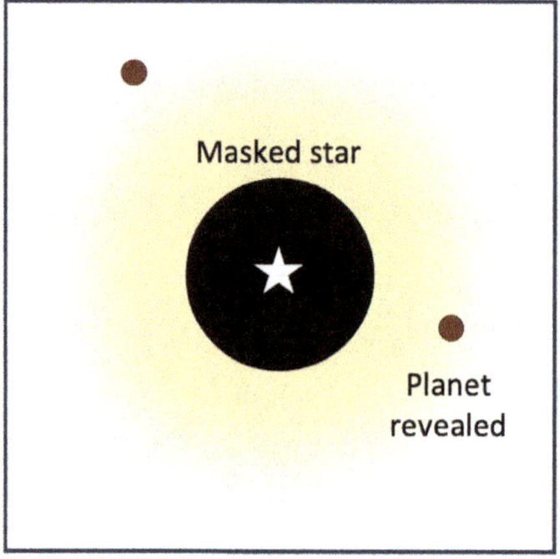

Figure 5.4

The First Discovery of Exoplanets

The existence of exoplanets was predicted a long time ago, in the early 1900s. However, the first confirmation of an exoplanet did not happen until 1992, and it was not what scientists expected. The first exoplanets officially discovered did not orbit a typical star; they orbited a pulsar. Yes, it is possible that after a star dies, the planets remain bound to the core and continue to orbit it. It is also possible for the dead star to capture other planets with its strong gravity. While monitoring the radio pulses from a pulsar, astronomers Wolszczan and Frail noticed that the pulses' timing was occasionally slightly misplaced. They concluded that this was due to two massive planets, which are 3.4 and 2.8 times the mass of the Earth, orbiting the pulsar. This was the first confirmed discovery of exoplanets, but it was orbiting a pulsar, not a star like the planets of the solar system that we are used to. This method of detection used pulsar timing, and it is specific to exoplanets orbiting pulsars (not mentioned in the paragraph above).

Technically, the Wolszczan and Frail discovery might not have been the first discovery of an exoplanet. In 1988, a team of Canadian astronomers detected wobbling from a star and suspected that this indicated the existence of the first exoplanet that they called Gamma Cephei Ab or Tadmor. It is worth mentioning that the name of exoplanets is composed of the name of their star, in addition to a letter designating the planet. In 1992, the team withdrew their claim because the quality of the data was not good enough to claim a detection. It turned out that in 2003 the planet's existence was confirmed by another team. The Nobel prize for the first detection of an exoplanet went to Michel Mayor and Didier Queloz, both of whom detected the first exoplanet orbiting a star, 51 Pegasi, while using the Haut-de-Provence observatory in France, as well as the wobble method. The planet was named 51 Pegasi b, and the discovery was announced in 1995. This was the first exoplanet discovered orbiting a star, like the planets we know in our system. Michel Mayor and Didier Queloz shared the 2019 Nobel prize for discoveries on the evolution of the Universe and Earth's place in it, with James Peebles, a Canadian-American astrophysicist who was working on cosmology.

Types of Exoplanets

Astronomers have classified exoplanets following our knowledge of planets in the solar system. The main categories are:

Gas Giants and Neptunians

Gas giants in our solar system, such as Jupiter and Saturn, are mainly composed of Hydrogen and Helium. Gas giants are the easiest exoplanets to find. Since they have large masses, they will cause their stars to wobble the most, making it easier for astronomers to detect this wobble. Moreover, their large size will cause more starlight dimming, which is also easy to detect.

What intrigued astronomers with the discovery of planet 51 Pegasi b, is that it was a gas giant more massive than Jupiter, orbiting very close to its star, at a distance of only 8 million kilometers; ~6 times closer than Mercury to the Sun. The fact that a gas giant could orbit so close to the star was difficult to comprehend at the time, because in our solar system, we are used to finding gas giants far from the Sun. For example, Jupiter is 100 times farther from its star at 746 million km. Due to its close distance to the star, 51 Pegasi b has a surface temperature between 500°C to 1,000°C. Such gas giants that are close to their stars are called hot Jupiters. The most massive exoplanets discovered to date are around 30 times the mass of Jupiter. They could be classified somewhere between gas giants and brown dwarfs. As you have probably figured out, Neptunian exoplanets are exoplanets of similar size to Neptune. Like gas giants, they are dominantly made of Hydrogen and Helium, and they have a rocky core. Neptunian exoplanets that are close to their stars are called hot Neptunes.

Super-Earth and Terrestrial

Super Earth and terrestrial exoplanets are rocky planets. Super Earth planets are rocky, gaseous or a combination of both. They are more massive than Earth but less massive than giant planets. There are no Super Earth-like planets in our solar system. These planets are the hardest to find because of their relatively small size. The easiest ones to find are the ones closest to their stars, as they cause a slight dimming of starlight when they pass in front of our line of sight.

Exomoons

Most planets in our solar system have moons. So why would planets in other systems not have any? The idea of exomoons dates several decades back and is not new at all. For example, an important part of the Star Wars movie, Episode 6, takes place on the Ewoks' habitat: a moon for the planet Endor. While the alien planet is a gas giant that cannot support life, its moon, the Forest Moon, is a habitable world. In reality, however, exomoons remain to be confirmed.

The first candidate exomoon came in 2017 from the Kepler Space Observatory. The candidate moon is Kepler 1625b I, a possible moon for the exoplanet Kepler-1625b. This was the first candidate exomoon, but it was not the only one. In the meantime, the candidate detections are awaiting confirmation.

The naming scheme of an exomoon is done by adding a Roman numeral after the planet's name. So the moon's name will have the star name, followed by the letter designating the planet, followed by a Roman numeral designating the moon. In the case of an exomoon, astronomers search for dimming light from exoplanets when moons pass in front of them, or they observe several starlight dips during the planet and moon transit. The discovery of exoplanets is already difficult enough with current technologies, and the discovery of exomoons is even more difficult.

There is a topic that I did not mention in Chapter 3: The possibility of a moon orbiting another. The existence of such objects is believed to be possible. After all, our Moon is large enough to have its potential natural satellite. The name of moons orbiting other moons is moonmoons. To date, no moonmoons have been discovered yet; not in our solar system or anywhere else in the Universe.

Some Interesting Exoplanets

There are more than 5,000 confirmed exoplanet discoveries to date. I cannot talk about each one specifically, but there are some strange worlds beyond what we can find in the solar system. Some of the planets mentioned previously in this chapter are already strange enough, like 51 Pegasi which orbits so close to its star with an orbital period of just 4.5 Earth days. The other strange one would be the first exoplanet discovered since it orbited a dead star. In this section, I will highlight a few exoplanets with strange features.

Kepler-16b

Kepler-16b is an exoplanet that orbits a binary star: two stars. I chose to talk about Kepler-16b since I like Star Wars, and it is a Tatooine-like planet. The sunset on Kepler-16b features two Suns setting down the horizon, just like on Tatooine.

WASP-76b

WASP-76b is an exoplanet that is tidally locked to its star. This means that the planet's revolution around the star and its rotation around itself have the same periods, resulting in one face of the planet permanently facing the star. Like in the case of the Moon's orbit around Earth. The exoplanet is just 0.033 au away from its star. This is why the temperature on the side facing the star reaches around 2,300°C. This is hot enough to vaporize iron. In fact, on this planet, it rains iron.

Kepler 10-b is another exoplanet with a similar scenario of the iron rain.

Another planet where the weather would be deadly strange is HD189733b. On this planet, it is so hot and windy (the wind can reach 7 times the speed of sound) that it rains molten (melted) glass, sideways.

55 Cancri e

55 Cancri e is another hot exoplanet that orbits so close to its star that its surface temperature can reach 2,300°C. 55 Cancri e is a rocky planet, a super-Earth. The intriguing fact about this world is that it has 8 times the mass of Earth. This led scientists to believe that the exoplanet could be composed of highly pressurized carbon, or in other words, diamonds. Imagine that somewhere out there, there is a planet made entirely of diamonds. With the current value of diamonds, this planet could be worth nonillions (30 zeros) of dollars, if this is confirmed. In comparison, the world's GDP is around a hundred trillion (12 zeros)

dollars. Our first intuition would be to go there and bring back all the diamonds we can get. However, firstly, this would be impossible with current technology, since the exoplanet is light-years away. Secondly, if we bring all these diamonds to Earth, diamonds might not be that valuable anymore. It may however be useful for some manufacturing industries where they use diamonds as cutting tools.

The First Exoplanet in Another Galaxy

All the exoplanets discovered until 2021 are in our galaxy, with a radius of less than 100,000 ly. In 2021 scientists may have found an exoplanet in another galaxy, more specifically in the whirlpool galaxy, 28 million light-years away from Earth. Even though the existence of exoplanets beyond our galaxy is already predicted, this potential discovery might bring the awaited proof.

⋮

Earth Analogs and the Hunt for Exo-life

In our search for exoplanets, we can seek an Earth analog that can potentially host life. While Earth 2.0 is not yet found, several exoplanets have been discovered similar to Earth and may be potentially good candidates to host life. These are usually rocky planets, around the same size as Earth, and orbit around their stars in the Goldilocks' zone. Their existence in the Goldilocks' zone indicates that the temperature on the planet may be suitable to have liquid water, which is an essential element for life as we know it.

Many similar planets have been found. For example, TOI-715 B is a super-Earth in the habitable zone of a M-type star, with a distance of around 0.083 au from its star. The orbital period of this planet is 19.3 days. One year for this planet is equivalent to 19.3 Earth days. Although it is at a much closer distance from its star than Earth, it is in the habitable zone, since its star is cooler than the Sun. So the zone where the temperature is suitable to have liquid water is closer to the star. This planet might have an important requirement to host life, such as a moderate temperature, but it does not mean it is currently suitable for life as we know it. Planet, for example, may suffer from a critical greenhouse effect like Venus, which can raise the temperature on the surface to hundreds of degrees. The habitable zone of a star depends on the temperature and size of the star. Not all habitable zones are similar to the sun's. Some might be farther if the star is bigger and hotter, and some might be closer if the star is smaller and cooler. In the TRAPPIST-1star system, which is 40 ly away from us, 7 exoplanets were discovered where 3 or 4 of these planets are believed to be in the star's habitable zone. When searching for life outside the solar system, we usually look for forms similar to life on Earth since it is the only form of life we know well. But this also means that other forms of life might exist out there. These exoplanets are light years away from Earth, meaning that if we want to check the planet out and search for life, it will take us thousands to millions of years to reach them with current technology. However, this does not mean there isn't something we can do to check for the possibility of life. A good indicator that a planet can host life is the existence of oxygen in its atmosphere, O_2 in particular.

Not only does life need O2, but life itself can produce O2 in abundance. On Earth, most of the O2 in the atmosphere is produced by organisms such as plants and algae. These organisms ensure the abundance of O2 in the atmosphere and its renewal.

So, when searching for life in exoplanets, we need to look at which exoplanets have abundant oxygen, similar to the levels on Earth. When a planet has oxygen, like Earth, it may indicate that it could be a potential home for life. Luckily, we don't have to go there to check for oxygen levels. It will suffice to look at the planet's atmosphere and identify the elements present through spectroscopy. To do so, powerful telescopes are needed. Space agencies are developing such instruments to probe the possibility of life in the Universe.

Our Place in the Universe

The discovery of exoplanets has changed our understanding of astrophysics in general, and many of our previous beliefs were altered by these discoveries. For example, the existence of gaseous planets close to their star is new to us. Previously, we mistakenly thought that gas planets usually formed and stayed in the outer regions of a system. This misconception changed our understanding of the solar system's evolution and history and forced us to think of new theories that could explain the existence of gas planets in our solar system, far from their star. In other systems, they are very close. The discovery of new types of planets broadened our horizons. For centuries, our understanding of planets came from our few planetary neighbors. Today, we have thousands of examples to study and learn from. The immense added diversity also showed us our place in the Universe. What we have been observing for centuries is only specific to one dot in the Universe. However, the Universe has many more wonders to offer.

The discovery of exoplanets has led to a major impact on our search for extraterrestrial life. Can you imagine how much the odds of finding extraterrestrial life have increased? Instead of looking for life in eight planets and their moons, we are now searching billions and billions of planets, and billions and billions of moons. Moreover, we have learned that we are made of the same elements of the Universe; we are not made of unique elements. In the words of Neil DeGrasse Tyson, what makes us special is that we are the same. We live in the Universe, and the Universe lives in us.

I don't want to turn this book into a philosophical one, but I would like to invite you to reflect, for a few seconds, on humanity's place in the Universe and the possibility of extraterrestrial life. I think that both options, the existence and absence of extraterrestrial life, would have a deep impact on my own considerations. If extraterrestrial life exists, I would be delighted and I would want to learn about it as much as possible. Whether it is intelligent life or just some sort of bacterial form, it would mean that we are not alone in this vast Universe. The diversity of species is not bound to one planet anymore, but it is intrinsic to the Universe itself. This should encourage us to reconsider our identity, ego, territorial, and race-based quarrels on Earth. We would no longer be children of nations; we would become children of Earth. However, if it turns out that life is bound to Earth and that there is no life out there, it would be unfortunate for my curiosity. It might seem ironic to have been given an immense Universe where life appears insignificant to it, yet this could imply that we might be the finest products of life. This drastically increases our responsibility to truly be at our greatest, and to make the best out of an entire Universe.

Chapter 6

The Milky Way and the Local Group

Andromeda
Triangulum
Halo
Disk
Bulge
Nucleus
Milky Way
Large & Small
Magellanic
Clouds

A galaxy is a group of stars, dust, gas, and other things gravitationally bound together. There are trillions of galaxies in the Universe, and our galaxy is the Milky Way where our solar system lies. The Milky Way is estimated to contain hundreds of billions of stars like the Sun. We can establish an analogy between a galaxy and a sand beach. If a galaxy is a beach, then a star system would be a grain of sand on that beach. The Milky Way would be a beach patch, and the Sun and the solar system would be tiny grains of sand on it. In this chapter, you will start to notice a significant change in the scales in space. We will shift entirely from the kilometers and astronomical units to the light year(ly) scale to describe distances. As mentioned at the beginning of this book, one light year (ly) is the distance that light travels in a vacuum during one year, equivalent to 9.5 trillion kilometers. In comparison, an astronomical unit (au) that depicts distances in the solar system is 150 million kilometers; 63 thousand times smaller than a light year. In this chapter, I will discuss our galaxy and its neighborhood. In the next one, I describe several types of galaxies found in the Universe.

The Milky Way

The Milky Way is a spiral galaxy. It appears to be a flattened disk, with several spiral arms of stars, dust, and gas, with a 100,000 ly diameter. The Milky Way contains between 100 to 400 billion stars; some are similar to the Sun, some are bigger, and some are smaller. The mass of the Milky Way is about 1 to 2 trillion times the mass of the Sun. A new study suggests that these numbers are actually 5 times less, but for the rest of this book we will consider the mass of the Milky Way to be 1.5 trillion Suns. The Milky Way revolves around itself at different speeds depending on how far we are from the center. For example, the Sun is at around two-thirds of the distance to the center, and it completes an entire revolution around the center every ~226 million years. The Milky Way comprises a nucleus, a central bulge, a disk, and a halo.

The Nucleus

Like every other spiral galaxy, the Milky Way has a center, and in this center lies a supermassive black hole. We learned about black holes in Chapter 3. Supermassive black holes have the same description as black holes, except they are 100,000 to several billion times more massive. The supermassive black hole in the center of the Milky Way is called Sagittarius A*. It is 4 million times the mass of the Sun. Like regular black holes, supermassive black holes do not emit light. So how did we discover Sagittarius A*?

It all began in 1931 when Karl Jansky, an American physicist and engineer, found a radio signal coming from the Sagittarius constellation toward the center of our galaxy. This radio source was later associated with the supermassive black hole in the center of our galaxy. The most compelling evidence we got about the existence of a supermassive black hole in the center of our galaxy came much later. In the early 2000s, astronomers studied the motion of a star called S2 near the radio source in the galactic center. Reinhard Genzel and Andrea Ghez led two independent studies in which the motion of the star was observed over a period of 10 years and discovered that the star was orbiting a massive invisible object. Their calculations showed that this hidden object should have a mass of several million solar masses. This fits well with the theoretical description of a supermassive black hole, a huge mass in a small portion of space. Later observations, based on the motion of several stars in the galactic center and observation of Sagittarius A* in different wavelengths, confirmed this discovery. Genzel and Ghwez were awarded the 2020 Nobel Prize for "supermassive compact object at the center of our galaxy" (shared with Roger Penrose). Note that this is the mention of a compact object, not a black hole specifically.

In one of the conferences, I attended in Prague (Texas Symposium), there was a discussion on the nature of the object based on the statement "compact object" instead of "black hole". A response that I liked was a mention of the quote, "If it looks like a duck, swims like a duck, and quacks like a duck, then it probably is a duck." It is established now that the center of the Milky Way is occupied by a supermassive black hole driving the rotation of the galaxy. Moreover, it is established today that all massive galaxies have a supermassive black hole at their cores. In 2022, the Event Horizon Telescope (EHT) took a snapshot of the Milky Way's supermassive black hole, but more on that in Chapter 11.

The Galactic Bulge

The galactic bulge is a region around the galactic center that is very dense with stars. From the outside, it is the brightest-appearing region of a galaxy. The stars in that region are usually old as they're the first to be formed. The galactic bulge also contains some gas and dust. The Milky Way's bulge is around 10,000 ly wide.

The Galactic Disk

The galactic disk is a spiral distribution of stars in the galactic plane centered on the galaxy's center. The Milky Way contains four main spiral arms: The Norma and Cygnus arm, the Sagittarius arm, the Scutum-Crux arm, and the Perseus arm, alongside other smaller arms. The arms are where most star formation occurs, and it hosts the newest stars in the galaxy. This is because the arms are rich in gas; the main ingredient for star formation. The arms rotate around the center of the Milky Way. In a nutshell, the rotation induces gas compression in the arms, bringing the gas molecules together and forming stars, as described in Chapter 3. The star population found in the arms is generally younger than the population in the center.

The galactic disk is divided into two regions, a thin and a thick disk.

Dust, gas, and younger stars are often found in the thin disk. It is only a few (around four) hundred light-years wide but contains 85% of the stars in the galaxy. The thick disk, on the other hand, is a few (one or two) thousand light-years wide. The galactic disk, alongside all the other galaxy components, rotates around the galaxy's center. The rotation speed varies depending on the location of the disk.

The Halo

The halo of the Milky Way is the spherical region around it. The Milky Way's halo extends to more than 20 times the size of the Milky Way. The stellar halo is composed of stars, mainly old stars, and globular clusters of stars. The halo also contains gas and dark matter (dark matter halo) that extend to regions farther than stars. We will talk about dark matter in the next chapter. What is important to know here is that dark matter composes most of the mass of a galaxy, but let's keep this suspense for the next chapter.

.
.
●

The Interstellar Medium

The Milky Way is the home to 200 to 400 billion stars. These stars are very far away from each other, and in between them, there is a lot of space. For example, Proxima Centauri is the closest star to the Sun: 4.25 light-years away. A probe traveling from Earth to the Sun would take a few dozen years to reach the solar system's edge. In contrast, getting to Proxima Centauri would take several 10,000 years. The Northern Star is about 323 light years away. Other close stars we see at night are hundreds of light years away. The farthest stars on the other edge of the galaxy are tens of thousands of light-years away.

The space between stars is not empty. As mentioned before, a galaxy contains gas, dust, and stars, and interstellar space is filled with gas and dust. Some regions are denser than others, especially in the arms. Among the most beautiful objects we can find in the interstellar medium are planetary nebulae. These are the relics of stars, the leftovers of stellar deaths. These nebulae are made of gas and dust scattered by the stars during their life and death, and they can take several shapes. They are often named after the objects they resemble. Some known nebulae are the Crab, the Helix (the closest), Orion, the Ring, the Cat's eye, the Butterfly, the Skull, the Oyster, and the Dumbbell nebulae. Large nebulae can be stellar nurseries where many stars are born from gas and dust.

Earth in the Milky Way

The solar system is in the Orion arm of the Milky Way, in the thin disk. It is at a distance of 26,000 ly from the center. At the solar system's position, the revolution period around the galaxy's center is approximately 226 million years. The Sun takes 226 million years to complete a circle around the galaxy's center. Since the extinction of the dinosaurs, the Sun has only completed one-quarter of a revolution around the center of the galaxy. The last time the Sun was in this position, relative to the galaxy's center, Pangea, the supercontinent still existed on Earth. Less than a quarter rotation later, the landmasses began to separate to form the seven continents we know today.

The solar system is tilted with respect to the galactic plane. This means that as the Sun orbits around the Milky Way, the solar system and galaxy planes are not aligned, but they form an angle of 60° instead. When you are away from city lights, in a location like the desert or the mountains, and you look at the sky, you will notice a long whitish trail across the sky. This trail is our galaxy as we look at it from the inside, and the whitish color is due to the light of the numerous stars. We are in the belly of the beast (the Milky Way), looking at it from the inside out. This is where the name of our galaxy, the Milky Way, comes from. If you happen to be in the southern hemisphere, you will notice a concentration of this whitish trail; this is the galactic center.

Because of the Earth and the Sun's tilt, with respect to the galaxy, the galactic center appears to be in the southern hemisphere and not at the equator. It is very difficult to get a real image of our galaxy as we are inside of it. Instead, we try to compute an image of the galaxy by mapping the stars, gas, and dust in it or by using simulations.

The Local Group

After we have been introduced to our home, the Milky Way, and found our place in it, let's look at our neighborhood. The word neighborhood implies much greater distances now. As you might have suspected, the Milky Way is not the only galaxy in the Universe. There are thousands of billions of other galaxies in the Universe, but we are part of what we call the local group. It is a group of a few galaxies with three major galaxies; the Milky Way is one of them. The second is Andromeda, the largest one in the group, roughly twice the size of the Milky Way, and is 2 million light-years away. The third galaxy is the Triangulum galaxy, which is smaller than the Milky Way and the third largest of the group. The Andromeda galaxy can be seen by the naked eye as a cloudy dark shape in the night sky. The local group size is about 10 million light years and is centered somewhere between the Milky Way and Andromeda.

The group also contains over 30 small galaxies. From these small galaxies, there are dwarf galaxies that are associated either with the

Milky Way or Andromeda as satellites. The most famous dwarf galaxies associated with the Milky Way are the Small and Large Magellanic clouds, seen by the naked eye from the southern hemisphere. Some of the Milky Way's satellite galaxies are torn apart by our galaxy's gravitational pull. Leftover trails of stars, gas, and dust from these dwarf galaxies can be seen inside and around the Milky Way.

The local group will witness a major event in 5 billion years, because the Milky Way and Andromeda, the two massive galaxies of the group, are pulling toward each other and will eventually collide in 5 billion years. The two galaxies will then form a new enormous galaxy. This might seem like a cataclysmic event, however, due to the large gaps between stars in a galaxy, the chance of any stars colliding during this fusion is very slim. This impending collision will resemble a peaceful galaxy waltz more than a destructive event.

Extragalactic world

In this chapter, I will talk about galaxies. Galaxies are the building blocks of the Universe. I will take you on a journey through the Universe, describing its vastness on the largest scale. Together, we will explore scales that range from thousands to several billion light years. In the end, I will discuss the first of two dark components in our Universe: dark matter and dark energy. Dark matter and dark energy are poorly understood today; however, they compose around 95% of our Universe and are primarily dominant in the Universe's overall energy budget. Dark matter will be featured here, and dark energy will be tackled in the next chapter.

Galaxies

In the early 20th century, the existence of galaxies outside the Milky Way was highly debated. At this time, observations were less sophisticated, and potential galaxies were called nebula because they resembled clouds of gas in the sky. Astronomers debated whether these nebulae were inside or outside the galaxy. It was not until 1923 that Hubble proved that these nebulae are outside the galaxy, and they are galaxies on their own.

Today we know that the Universe is made of an estimated 200 to 2,000 billion galaxies (but these numbers can change). These galaxies have different masses, sizes, and shapes, ranging from tiny galaxies, called dwarf galaxies, to giant galaxies. The Milky Way, our home, is just one of them. From a morphological point of view, galaxies are classified into three main types: spiral, elliptical, and irregular. The Hubble sequence, often referred to in astrophysics, represents the major types of galaxies. The Milky Way, Andromeda, and Triangulum galaxies are spiral galaxies.

Spiral Galaxies

As their name indicates, spiral galaxies appear to be flattened galaxies with a spiral shape. The previous chapter gave a first-order description of the Milky Way. The Milky Way is a spiral galaxy; its components (nucleus, bulge, disks, and halo) are generally found in all other spiral galaxies. There is another component that I did not describe in the previous chapter: spiral bars. Spiral bars are central bars in the galaxy that can link the arms together. Spiral bars are rich in gas, and are believed to be stellar nurseries, as star births are frequent in this region. Around two-thirds of spiral galaxies are barred spiral galaxies. Spiral galaxies are usually rich in gas and are home to star formation.

A helpful rule to remember is that where there is gas, there is probably star formation. Since they are actively forming stars, the star population in spiral galaxies is mainly young. At the beginning of the life of a galaxy, a disk is formed, similar to when stars are formed. Stars are born in this disk. Then when other small galaxies pass by, their gravitational pull creates a swirl in the disk and forms the spirals. Spiral galaxies are usually young active star-forming galaxies. Spiral galaxies are bright and easy to spot. This is why they may be easier to detect. In the nearby Universe, 60% of galaxies observed are spirals. Spiral galaxies have a mass ranging from a few billion to a few thousand billion solar masses.

The Milky Way itself is considered a large spiral galaxy; it has a mass of 1,500 billion solar masses. NGC 6872 is one of the largest spiral galaxies discovered, and it is 5 times bigger than our Milky Way (500,000 ly across). It contains between 500 billion to 2 trillion stars. UGC 2885 is another spiral galaxy that contains 10 times as many stars as the Milky Way. It is believed to be between 460 and 800 thousand light-years across and contains 10 times more stars than the Milky Way. However,

UGC 2885's halo is only believed to be that large, while the size of its spirals is believed to be 2.5 times the size of the Milky Way, making it less than 300,000 ly across.

Elliptical and Lenticular Galaxies

Elliptical galaxies, as their name indicates, have elliptical or spheroidal shapes. They don't have any particular features like spiral galaxies. They are mainly formed of old stars, contain very little gas and dust, and do not have an important star formation activity. One way elliptical galaxies are formed is by the merging of galaxies. Another way a galaxy becomes elliptical is when it consumes all its gas until it becomes an inactive, dormant elliptical galaxy. In other words, the star-forming factories in elliptical galaxies are partially shut down, making elliptical galaxies a home for old stars that have lived for billions of years. The only stars that can live that long are low-mass stars, which means that elliptical galaxies are composed of low-mass stars, hence they will have lower brightness than spiral galaxies. Therefore, they might be dimmer and more difficult to detect.

In the nearby Universe, 20% of galaxies observed are elliptical. Elliptical galaxies have a vast size and mass range, unlike any other galaxy type. Some elliptical galaxies can have masses equivalent to a fraction of the mass of the Milky Way, while others can be a thousand times more massive. Remember, the Milky Way is considered one of the big spiral galaxies. Lenticular galaxies are a close cousin to elliptical galaxies. They have a lenticular shape, lens-like appearance, and have a bulge and a disk component. They are a cross between spiral and elliptical galaxies The largest galaxy found to this day is IC 1101, an elliptical galaxy that stretches to 4 million light-years across and contains 100 trillion stars.

Irregular Galaxies

Irregular galaxies do not have any distinct shape. They are chaotic in appearance and do not seem to contain any features, like a central bulge or arms. Irregular galaxies are very rich in gas and they form stars. This is not necessarily the case for dwarf irregular galaxies, which are tiny irregular galaxies like the Small Magellanic Cloud. Compared to spiral galaxies, irregular galaxies have a larger range of star formation rates. Some of the irregular galaxies form stars at very high rates and are called starburst galaxies. Several channels exist for irregular galaxy formation, the most common being galaxy interactions. In other words, irregular galaxies form when galaxies (spiral or elliptical) merge, or when a galaxy passes by another one, causing it to deform under its gravitational pull. Irregular galaxies are most common in large groups of galaxies, where galaxy interactions are frequent.

Irregular galaxies do not show any features in general, however some of them still contain characteristics of the old galaxies before they were deformed. For example, the Cartwheel galaxy appears to be formed by a small galaxy passing in the middle of a larger one. The features of the two galaxies can be seen clearly, with the larger galaxy forming a large wheel and the smaller galaxy forming a smaller wheel in the center. Irregular galaxies are usually small, with masses of only a fraction of the mass of the Milky Way. The smallest ones, dwarf irregulars, can still contain a few hundred million to a few billion stars.

In the nearby Universe, 20% of the galaxies observed are irregular. Still, their universal mass budget in the Universe is small since they are low in mass. The Small and Large Magellanic clouds are one of the closest irregular galaxies to the Milky Way. They can be seen by the naked eye in the southern hemisphere. They were considered the nearest galaxies to the Milky Way until the discovery of closer galaxies, the Sagittarius Dwarf Elliptical Galaxy in 1994, and the Canis Major Dwarf Galaxy in 2003. The latter now holds the title.

Hubble Sequence

To classify galaxies according to their morphological characteristics, astronomers often use the Hubble sequence, published by Edwin Hubble in 1926. The sequence resembles a tuning fork and is illustrated in Figure 7.1.

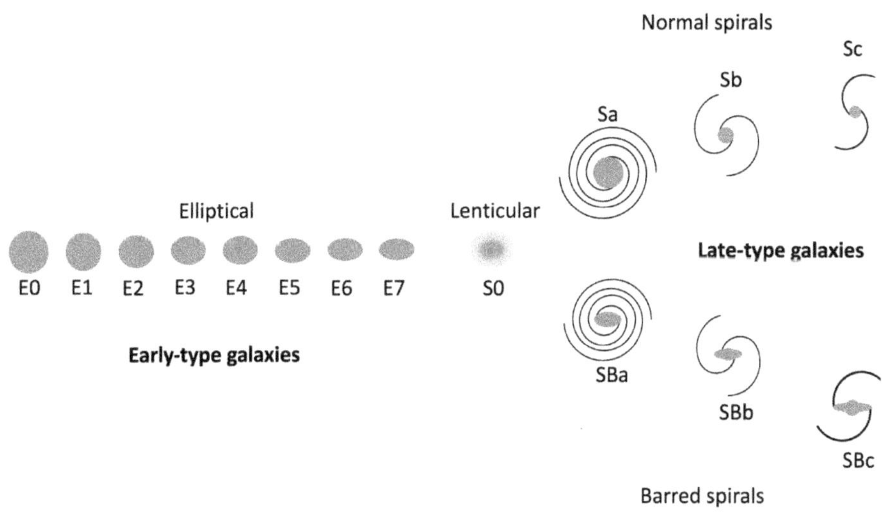

Figure 7.1

The first group in the Sequence is the Early-type galaxies, which have a spheroidal or ellipsoidal shape and include Elliptical galaxies. They are called Early-type because they used their gas to form stars very early in their lives.

The other group is the Late-type galaxies that constitute Spirals and Spiral-barred galaxies. These galaxies are divided into sub-groups following their spiral shape. They are called Late-type because they did not form stars early in their lives.

A fourth class of irregular galaxies was added. Lenticular galaxies are transitional galaxies, between elliptical and spiral, and can be found in the intersection between Early and Late-type galaxies. The Hubble Sequence was originally intended to describe the evolution of galaxies from Early-type to Late-type, but it turned out that the early Universe was more abundant with spirals and irregular galaxies.

Active Galactic Nuclei

Some galaxies have a very bright and active core. These cores are called Active Galactic Nuclei (AGN), and they are the brightest objects in the Universe. The activity of AGNs is caused by gas and dust falling in the central engines of galaxies: the central supermassive black hole. This causes the black hole to launch a deadly jet of high-energy particles and rays, extending to several hundred light years. The launched jets contain harmful gamma rays that can wipe out entire civilizations. However, rest assured, there are no AGNs dangerously close to Earth.

The terminology of these bright objects depends on the angle at which the jet is seen from Earth. If the jet is directed toward Earth, the AGN is called a Blazar. If the jet is at an angle perpendicular to the line of view from an observer on Earth, it is called a Quasar. Blazars and quasars are the same objects showing different emission features because they are seen from different angles. AGNs can be found in all types of galaxies and are believed to play a significant role in galaxy evolution. They can affect star formation in the Universe. For example, in one of the scenarios of the formation of elliptical galaxies, it is believed that the supermassive black holes in the center of the spiral galaxies can become very active after merging with another galaxy. It will consume all the gas and dust in the galaxy until it becomes a dormant elliptical galaxy. One instance where a positive impact of AGNs on the formation of stars can be observed is in the elliptical galaxy M87. M87 has an AGN in its core shooting out jets of matter and radiations to a distance of more than 5,000 ly. The pressure waves caused by these radiations compress the gas at the edges of the M87 jet, causing high-density gas regions in the galaxy. M87 is an inactive elliptical galaxy. However, the regions at the edges of the jet are star-forming regions due to the creation of gas-dense regions by the jet. This is an example of how an old elliptical low-activity galaxy can have local regions of high star formation rates.

The Cosmic Web

We now know the major types of galaxies, a rough approximation of their numbers, and their main characteristics. In this paragraph, we will tackle how these galaxies are distributed in the Universe, and how the matter in the Universe is structured. The structure of the Universe at the largest known scale resembles a spider web. In the next section, you will see why. The structure is composed of filaments and knots at the intersections. It is usually referred to as the cosmic web. The filaments are composed of galaxies with large concentrations at their intersections. Therefore, galaxies are either inside the filaments or clustered at the intersections. These intersections are called galaxy clusters. Galaxy clusters are the largest building blocks of our known

Universe. They are either regular clusters or superclusters. Each galaxy cluster can contain hundreds to thousands of galaxies and large amounts of intergalactic gas. Usually, in the center of each cluster, a giant galaxy is found, called the Brightest Cluster Galaxy (BCG). This galaxy is the heaviest in the cluster, and all other galaxies are grouped around it; they are gravitationally bound to it. If galaxy clusters can be considered bee hives, then BCGs are the queens. Stars only make up a few percent of the mass of clusters and filaments, while gas can contribute several times more to their mass budget. More than 80% of the mass comes from dark matter (To be explained in the next section).

The nearest cluster to the Milky Way is the Virgo cluster, nearly ~50 million light-years away. The Milky Way is believed to be at the edge of the Virgo cluster; we are not inside the cluster. The Virgo cluster is several million light-years wide and contains up to 2,000 galaxies. The central galaxy of the Virgo Cluster is M87; discussed in the previous section. M87 is an elliptical galaxy, a massive beast that is twice as massive as the Milky Way. Its star population is old, and its star formation rate is low. However, as mentioned before, M87 shows some star-forming regions due to its AGN. The Virgo cluster is the center of a larger entity, called the Local Supercluster, or the Virgo Supercluster. It is a large entity, stretching over more than 100 million light-years wide. It contains the Virgo cluster, the Local Group (which includes the Milky Way), and over a hundred other groups of galaxies.

It is estimated that the Universe contains 10 million superclusters.

M87 is not the most massive galaxy in the Universe, and the Virgo cluster is not the most giant cluster. The Coma cluster is a nearby cluster, 320 million light-years away. The cluster is 20 million light-years wide, contains thousands of galaxies, and is one of two main clusters of the Coma supercluster (the Leo cluster being the second one). The largest galaxy discovered until now is the lenticular IC 1101 situated in the Abell 2029 galaxy cluster. The largest galaxy cluster discovered to date is the El Gordo (the Fat one) galaxy cluster which is 3 quadrillion times more massive than the Sun, or 2 to 3 times the mass of the Virgo cluster. Due to their sizeable gravitational potential, galaxy clusters attract galaxies and gas from the filaments toward them. Therefore, filament galaxies appear to be traveling toward the center of the clusters. Since traveling toward the cluster takes time, most galaxies found in clusters are old galaxies; therefore, they are primarily elliptical. Galaxies of the filaments are usually young spirals.

In 2014, a new way of defining a supercluster emerged based on the relative velocities of galaxies. A new larger supercluster home of the Milky Way, called Laniakea, was identified in that way. It encompasses around 100,000 galaxies, including the Virgo supercluster and others, and stretches to more than 500 million light years.

The largest structure found in the Universe is the Hercules-Corona Borealis Great Wall (or just the Great Wall), which is a filament of galaxies that extend to 10 billion light years, or around 9% of the size of the observable Universe (the observable Universe is defined in the next chapter). The discovery of this structure was made through the spatial distribution of gamma-ray bursts, which I will explain in Chapter 9. Another noticeable filament-like structure is the Sloan Great Wall 1.4 billion light-years wide.

In summary, we have the cosmic web composed of superclusters, clusters, and filaments, and inside these lie hundreds to thousands of galaxies. Those comprise billions of stars, gas, dust, and dark matter. Around stars orbit planets, dwarf planets, asteroids, comets, and many small objects. Some moons orbit around some of those planets. Some moons might also have other moons. Earth (with its Moon) is just 1 of 8 planets orbiting an average star. 1 star, of 400 billion other stars, in a galaxy sharing a supercluster with 100,000 more galaxies, which is part of a Universe with trillions of other galaxies, comprised of superclusters, clusters, and filaments.

Dark Matter

Until now, we know that galaxies are composed of stars (and their systems), gas, and dust. In the Milky Way, the mass of the stars, gas and dust only comprises less than 20% to one-third of the mass of the galaxy. The exact values can vary depending on the galaxy radius/region considered and the method of measurement. The important thing to know is that neither stars, gas, nor dust are the most massive components in a galaxy. The most massive component of a galaxy is dark matter.

So, what is dark matter, and how was it discovered? The first evidence of dark matter came in 1933 when astronomer Fritz Zwicky was observing galaxy clusters. He found that the mass needed to gravitationally bind the galaxies in the cluster, and prevent them from drifting away was much larger than the total mass of the individual galaxies observed in the group. This was the first evidence that something very massive had to contribute to the cluster's gravity. In the 1970s, Vera Rubin was drawing her path in astronomy in a society that was, at the time, not the friendliest to women in science. In her research career, she chose to study the rotation of galaxies, and she started observing spiral galaxies with her colleague astronomer, Kent Ford. In a spiral galaxy, the stars in the galaxy center supposedly move faster because there is a more significant concentration of matter. In contrast, at the edge, they supposedly move slower. This is similar to what we observe in our solar system with the rotation of the planets. However, this is not at all what Rubin observed. She observed tens of galaxies with the same results: it seemed that the stars at the edges of the galaxies had the same rotational speed as the stars close to the center. Rubin's calculations showed that to explain her observations, the mass in the galaxies should be 10 times greater than the visible mass in order to provide enough gravity to sustain its rotation. This means there is a type of unseen matter in the galaxies: dark matter.

At first, astronomers did not easily accept the idea of dark matter, however, further evidence and observations made it compelling. Moreover, the possible existence of dark matter is also seen in studies in other fields, like cosmology. So, what is dark matter? Well, all we know for sure is that dark matter does not appear to interact with anything and has no detectable emission. It is invisible to us, and we can only see it indirectly through its gravitational effect. We also know that over ~80% of the matter in the Universe is dark matter. We think that dark matter might have shaped the cosmic web as we see it today, through its gravitational effects.

The galaxies are submerged in the gravitational well of dark matter. After the Universe started expanding and cooling down, dark matter, under gravity's effects, started clumping together. Ordinary matter was attracted to these clumps and gathered to form stars and galaxies. The flow of dark matter from low to high-density regions formed the filaments in the cosmic web. Dark matter seems to be located around galaxies and other concentrations of regular matter. This little information on dark matter might feel like a deception for the reader who is eager to learn what dark matter is. But rest assured, we are all keen to know more about it.

Today hundreds of theoreticians and experimentalists are trying to understand dark matter. Some suggest that dark matter might be a new type of particle with very little interaction with its surroundings. These particles are referred to as Weakly Interacting Massive Particles (WIMPs). We are searching for WIMPS directly in the labs, and indirectly in space (mainly looking for their possible interactions and byproducts). Others suggest that the effects of dark matter might be due to hidden matter that we are familiar with, like primordial black holes. On the other hand, some scientists are working on alternative theories of gravity that could explain observations suggesting that our understanding of gravity or that our measurements are flawed or incomplete. One thing is sure until now: no dark matter candidates or alternative theories of gravity have been confirmed yet.

Chapter 8

Cosmology

Until now, I've been talking about objects in space and the building blocks of the Universe. In this chapter, I will talk about the Universe as a whole: from its beginning to its evolution, and its possible end. I will guide you through the main topics in cosmology.

The Big Bang

You have already heard of gas, intergalactic gas, dust, matter, dark matter, galaxies, stars, planets, moons, asteroids, etc. So, where does it all come from? What is the origin of the Universe? The best theory today explaining the origin of the Universe is the Big Bang Theory.

The Big Bang happened around 13.8 billion years ago. It is believed that the entire Universe, including all the matter, space, and time was once contained in an infinitely small bubble, possibly smaller than a subatomic particle, or infinitely small. Suddenly this bubble 'exploded', and this Universe became denser and hotter than ever. Within 10^{-36} seconds to 10^{-32} seconds after this singularity, the Universe was in an inflation phase where it expanded drastically; then, the expansion continued but at a slower rate. The Universe was growing, and within a fraction of a second, it reached billions of kilometers in size. As it expanded, the Universe, initially extremely hot and dense, started to cool down. Within the first second of the Universe's lifetime, the first fundamental particles of matter and anti-matter started forming. Matter and anti-matter annihilated each other, but for reasons that nobody is sure of, some matter particles survived. The first protons and neutrons formed before the Universe were one second old. After a few minutes, the Universe's temperature dropped to less than a billion degrees Celsius. It was now cool enough to form the first atomic nuclei.

After 380,000 years, the Universe cooled down enough for these nuclei to capture electrons and form the first atoms: hydrogen and helium. Hydrogen is the lowest mass atom found: it consists of only one proton and one electron, and is the most abundant element. 380,000 years after the Big Bang, the Universe was filled with clouds of hydrogen and helium gas.

Now, you might ask: What was before the Big Bang? Where did all this matter, initially in this tiny bubble, come from? The answer is: well, nobody knows. However, some argue that the question itself is not correct. What was before the Big Bang? The question does not seem right since the Big Bang was the beginning; there was nothing before it. Time itself started with the Big Bang. Therefore, some argue that the logic of the question is flawed. An incorrect way of thinking of the Big Bang is to imagine a room with a tiny bubble in the middle that started expanding. This is not how it happened. A more accurate way to imagine it is that the entire room was the bubble, and that room started expanding. Space was also created by the Big Bang.

Cosmic Microwave Background

In 1956 the Bell Telephone Laboratories built a radio receiver. It was a horn-shaped antenna around 6 meters long. Two scientists, Arno Penzias, and Robert Wilson, used this radio receiver to study intergalactic medium. After eliminating all sources of background noise, they still found an annoying background noise in their data that they couldn't eliminate. This noise was uniform and came from all directions. They looked at every possible solution to explain the source of the noise, and they found that the noise was not due to any urban interference. They cooled down their instrument with liquid helium to eliminate any instrumental noise source, but the noise remained. They cleaned the antenna for hours and deployed pigeon traps, thinking the noise could be due to pigeon roosting, but the noise was still there. At the same time, at Princeton University, Robert Dicke was studying the origins of the Universe. He theorized that this origin should have left an imprint in the form of low-level radiations. The three scientists came together when Dicke visited Bell's labs and confirmed that the mysterious background noise was what he was looking for: The Cosmic Microwave Background (CMB). During the first 380,000 years after the Big Bang, the Universe was so dense that any radiation from the explosion was scattered by roaming particles. Electromagnetic emission was trapped inside this hot thick plasma soup. The soup started cooling down eventually, and 380,000 years after the Big Bang, as explained in the previous paragraph, it was cool enough for the first hydrogen atoms to form. Hydrogen atoms did not affect the radiation from the cosmic background; these radiations escaped. This point in time is referred to by the surface of the last scattering. After that point, the radiation was no longer trapped.

Wherever we look in the Universe now, this radiation is there. It is a faint radio emission, a 13.8-billion-year-old relic of the initial explosion that created our Universe. It traveled 13.8 billion years to reach us. It allows us to see how the Universe was 13.8 billion years ago, just 380,000 years after its birth. It was a young Universe, before the creation of the first stars, simple and empty of all the complex elements and objects we see today. The published results are proof of the existence of the Big Bang. After that, several missions were sent to space to study and map the CMB, with the latest being the COBE, then the Planck satellite that created the detailed maps of the CMB. When we look at the CMB, we look back at 380,000 years after the Big Bang. The era of photon-matter interaction ended with the surface of the last scattering, starting the recombination era. During this period, protons and electrons combined to form the first atoms marking the beginning of the atom formation age.

A blackbody is a body whose radiation depends only on its temperature. In that case, the CMB has a temperature of 2.726 K, or -270.424 °C. The COBE satellite detected anisotropies in the CMB everywhere in the sky. These anisotropies are very small fluctuations of temperature that indicate an inhomogeneous distribution of matter at the epoch of recombination. Going back in time, they point to differences in density in small and large-scale structures right after the Universe's very beginnings. These fluctuations reflect growing fluctuations of dark matter distribution creating denser and more massive regions. These growing regions would eventually attract regular matter by gravity, becoming seeds for galaxy growth and shaping the cosmic web (see Chapter 7). John C. Mather and George F. Smoot were awarded the 2006 Nobel Prize for their discovery of the blackbody form and anisotropy of the CMB radiation.

Today the CMB is the earliest thing we can observe from our Universe. It is the earliest observable evidence of the Big Bang. Before this time, any radiation was trapped in the plasma soup. Therefore, we are not able to see anything beyond that era, but this does not mean we will not eventually be able to observe what happened in the first 380,000 years. Even if the cosmic microwave background was trapped until 380,000 years later, it does not mean that other messengers could not escape the hot dense plasma soup. Such messengers are gravitational waves and cosmic neutrinos. With the emergence of new technologies that permit us to detect these kinds of messengers, many believe that we should be able, in the future, to detect these other relics from the early Universe.

·
·
●

The Expansion of the Universe

After the recombination era and the formation of the first atoms, Hydrogen and Helium atoms came together to form the first stars a few hundred million years after the Big Bang. In the first billion years after the Big Bang, the first galaxies were formed. When Edwin Hubble discovered galaxies for the first time, proving that the Milky Way was not the only Galaxy, and irreversibly changed our view of the scale of the Universe, he also made another equally important discovery.

Hubble found out that these distant galaxies were moving away from us. To his surprise, the farther the galaxy was, the faster it moved away. After several observations confirming this finding, he established the Hubble law, now known as the Hubble-Lemaître's law. This law relates the speed at which a galaxy travels away from us to its distance, using a constant known as the Hubble constant. We have a reasonable estimation of the value of this constant today (around 70 km/s/ Megaparsec); however, the exact value is still yet to be determined.

Redshift

For the interested reader, this brief paragraph explains how we can determine whether a galaxy is moving away or closer to us and how we can measure the distance of galaxies. When observing a galaxy moving toward or away from us, we witness an effect similar to hearing an ambulance passing by. When the ambulance is coming toward you, the sound of the siren has a high pitch; as the ambulance passes you and gets farther, the pitch becomes lower. This change in sound frequency is known as the Doppler effect, theorized by Christian Doppler in 1842. A similar phenomenon happens when we observe a galaxy, except this time, it is light we see that changes instead of sound. When a galaxy is moving toward us, its colors shift toward blue, and when it is moving away from us, its colors shift toward red. This is why we talk about redshift. The redshift is measured using spectroscopy. It is determined by comparing the position of a color or an absorption line in the spectrum to a reference measurement as shown in Figure 8.1. As established by Hubble, the further a galaxy is, the faster it moves away from us; therefore, the more redshift its light will be. So, by measuring the redshift of a galaxy, we can deduce the velocity at which it is moving away. Then using the Hubble law, we can infer its distance. On a side note, spectroscopy can also be used to study the motion of nearer objects, like planets' relative rotations. This is done by comparing the spectrum at different points, such as the poles and equator. In addition, spectroscopy can be used to study star kinetics.

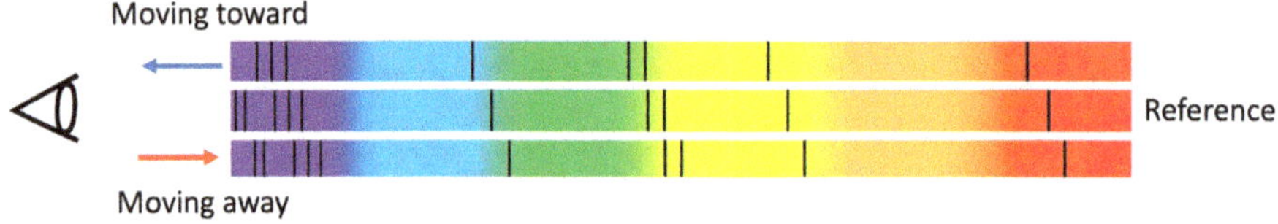

Figure 8.1

Dark Energy

At the same epoch as Hubble, Einstein was establishing the grounds of general relativity, a theory that could potentially describe the Universe at a macro scale. His set equation showed that the Universe is not static but in motion. Believing that the Universe is supposed to be static and that something was missing, he added a cosmological constant to his equation. After he was proved wrong by Hubble, he removed the constant from his equation and called it his biggest blunder. Logically speaking, if the Universe has been expanding since its formation, the expansion should be slowing down due to the effect of gravity. All the objects in the Universe are supposed to attract each other, slowing down the expansion of the Universe. Sounds logical, right? We could also speculate on the end of the Universe and predict that it will eventually collapse again. However, the Universe does not want to reveal its fate easily and seems much more complex than that.

In the 1990s, decades after Einstein and Hubble, astronomers discovered that not only are the galaxies moving away from us, but the Universe's expansion rate is also increasing. This can be explained by 'something' fueling and powering the Universe's expansion, and counteracting the gravitational pull that is supposed to slow the expansion. This 'something' was named: dark energy. Saul Perlmutter, Brian P. Schmidt and Adam G. Riess were awarded the 2011 Nobel Prize for the discovery of the accelerated expansion of the universe. If you ask me today what we know about dark energy, I would say that the only things we know for sure are the following: It is an energy from an unknown origin (dark being the right word here), that is causing the accelerated expansion of the Universe. We know that dark energy constitutes 70% of the entire energy budget of the Universe. In other words, today, we cannot explain where 70% of the energy in the Universe comes from. Several theories have been made to explain the origin of dark energy, but none are confirmed today.

In cosmological equations, the dark energy budget is accounted for in a constant similar to Einstein's cosmological constant, but with an effect opposite to what the cosmological constant of Einstein was supposed to do. Decades after the death of Einstein, the cosmological constant came back, and Einstein's biggest blunder did not seem like a blunder anymore; it looked more like the product of a wrongly interpreted ingenuity.

To summarize, after the Big Bang, there was an initial brief inflation phase that lasted a fraction of a second when the expansion of the Universe was at its peak. The expansion of the Universe continued after the inflation phase. It started to slow down a few billion years later under the gravitational effect of dark matter. Instead of collapsing a few billion years after that (around 8 billion years after the Big Bang), the expansion of the Universe accelerated again and is still accelerating today under the effect of dark energy.

Observable Universe

Our Universe is 13.8 billion years old. It is expanding, and this expansion is accelerating. As we observe regions in space that are farther away, we notice that galaxies are moving away from us increasingly faster. After a certain distance, the expansion of the Universe becomes faster than the speed of light. In modern physics, the speed of light is the speed limit that anything can reach. Nothing can go faster than the speed of light. All the laws of physics prohibit it. This law is valid for everything in the Universe, but not the Universe itself. Therefore, it is possible that at a farther distance from us, the expansion becomes faster than the speed of light. Don't think of the expansion as objects moving in space. Think of it as a new space being created between the different objects. Nothing can move faster than the speed of light in space, but space itself can expand faster. This means that the light emitted from galaxies can never reach us beyond a certain distance, since these galaxies are moving away from us faster than the speed of light. These parts of the Universe are unreachable and unobservable to us. This is why we talk about the observable Universe. The determination of the observable Universe does not consider our technological observational capabilities; rather, the laws of physics determine the observable Universe. It represents the Universe that we can see or would be able to see.

Our observable Universe is a spherical region centered around us that contains everything that we can see. This observable Universe contains 200 to 2,000 billion galaxies; the ones we were talking about earlier. It includes all the objects (galaxies) whose light can reach Earth. As for all the things beyond the observable Universe, their light will never be able to reach us at any point. When we look at a distant galaxy, we are looking at where the galaxy was when the light that we are seeing today was emitted. The actual position of this galaxy is supposedly farther away because it is moving away from us. The galaxy appears closer but is now far away from this observed position as shown in Figure 8.2.

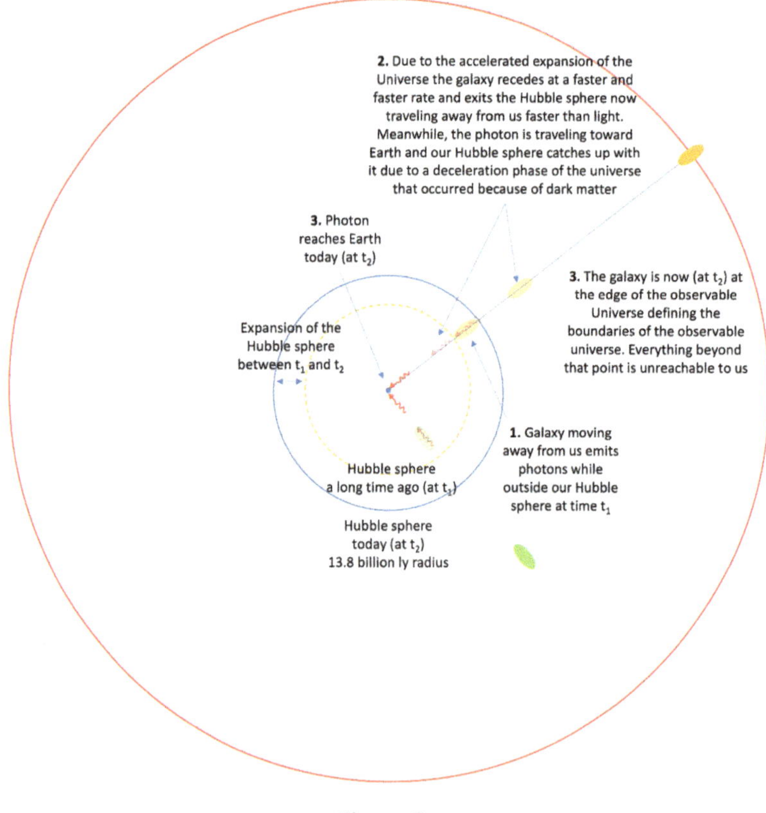

Figure 8.2

To calculate the size of the observable Universe, a common misconception would be to multiply the speed of light by the age of the Universe, which gives an answer of 13.8 billion light years. This number is the Hubble radius in cosmology, which is the radius of the Hubble sphere. As the Universe ages, our Hubble sphere expands as there is more time for light to travel. Galaxies at the edge of the Hubble sphere travel away from us at the speed of light. Beyond the Hubble sphere, they recede away from us faster than the speed of light. Consider a galaxy receding faster than the speed of light from our point of view. The emitted light at this point from this galaxy will never reach us. However, if this galaxy was inside our Hubble sphere in the past, traveling away from us slower than the speed of light, it is now beyond it due to the accelerated expansion of the Universe. As a result, light that it emitted while inside the Hubble sphere can still reach Earth. The galaxy is outside our Hubble sphere now, but we can see the light it emitted in the past. To calculate the size of the observable Universe, we should consider the actual position of the farthest galaxy we can observe, not its position at the time it emitted the light that we can see. This is why the actual size of the observable Universe is bigger than the Hubble sphere. Keep in mind that we are looking at this galaxy, as we do at any object, in the past. As the Universe ages, the Hubble sphere expands. As the Universe continues to expand, the observable Universe also expands. However, there will be regions that will not be observable to us anymore as they keep receding away from us at increasing speeds. At a certain point, they will travel away from us faster than light, and they will reach what we call the cosmological event horizon. It is unlikely that we will be able to observe objects beyond the observable Universe. Maybe, the answers to some of our mysteries lie there. Who knows...

Today's cosmology equations, which take into account the complex expansion of the Universe and its evolution, best estimate the size of the Universe to be around 93 billion light years (46.5-billion-light-year sphere radius). Considering that only signals emitted after the recombination epoch, including the emission of the Cosmological Background Light, can be detected from the early Universe, the calculated size becomes slightly smaller: around 91 billion light years. This is called the visible Universe. The observable Universe includes all the signals emitted since the beginning of the expansion. The visible Universe takes into account the fact that no signal could escape the hot dense Universe before the recombination period, and it includes only the signals emitted after this period. Our observable (or visible) Universe is determined by us, as we talk about the light that can reach us. Each observer is the center of its observable Universe. Ironically this makes us, in a way, the center of our (observable/visible) Universe.

•
•
●

Composition of the Universe

In cosmology, there is a standard model that describes the Universe and its composition called the Lambda-CDM model, where CDM stands for Cold Dark Matter, and Lambda is the cosmological constant associated with dark energy. In this standard model, the Universe comprises 68.3% dark energy, 26.8% dark matter, and 4.9% regular matter, also called baryonic matter. A very small fraction of it is composed of radiation and neutrinos. However, the estimation of these parameters can slightly differ with different measurements. Every object currently studied in the Universe, everything you have so far seen in this book, except dark matter and dark energy, constitutes less than 5% of our Universe. Most of this matter is composed of hydrogen and helium. More than 99% of the regular matter in the Universe is in the plasma state (different than the plasma in our blood). Plasma is a form of heated matter where the electrons are ripped away from the atoms forming ionized gas. It is much more abundant than matter in the solid, liquid, or gaseous state. In fact, most of the gas in the Universe, in stars, in the interstellar medium, or the intergalactic medium, is ionized. We don't know the nature of 95% of the rest of the Universe. In the standard model, dark matter is considered cold. This means that it cannot be dissipated and does not interact with other particles or with itself, except through gravity. In addition, it moves slowly. In other non-standard models, dark matter can be hot, and it moves very fast with velocities near the speed of light.

Data from several measurements, like the measurements of the CMB, support the Cold Dark Matter theory. There are sets of equations that describe the evolution and the composition of our Universe as it is today. The most common are the Friedman-Lemaitre equations, which are based on Einstein's general relativity, although they can take several forms and have many deviations. In these equations, a parameter (k) describes the curvature of the Universe, which can be either flat, curved, or hyperbolic. In the scenario where the Universe is curved, it would have enough gravity to overcome the force of expansion and curl into a ball-like shape. However, there is evidence today that suggests that the Universe is flat. Several experiments and cosmological measurements are being done to determine the exact parameters of the equations governing the physics of the Universe.

In our current understanding, there is a mixture of unknowns, many of which we cannot explain with our current understanding of physics. Dark matter and dark energy are not the only things we haven't fully grasped. For example, we don't know what is inside a black hole. By the way, black holes are a part of the 5% regular matter, and even within this 5%, there are still countless things that we don't understand. Scientists are looking at the skies to answer these questions by studying large-scale structures. However, to study the composition of the Universe, many are looking on the small scale, inside the atom, to solve these riddles.

Particle physics is one of the largest fields in physics, where scientists study the composition of matter in the Universe. They often do that by cracking atoms and particles open to see what they are composed of, and how they interact. One of the largest tools that is used is the Large Hadron Collider. It is based on the frontiers of France and Switzerland in CERN, the European Council for Nuclear Research (acronym derived from the French name Conseil Européen pour la Recherche Nucléaire). The Large Hadron Collider is a 27 km ring where particles are accelerated to nearly the speed of light and smashed into each other. This is useful for scientists to study the compositions of the particles in our world and its governing forces. Today, there are four forces we know of. The first three are the electromagnetic, weak, and strong forces, which reveal themselves on the atomic level. The fourth force is the force of gravity. Gravity is the weakest force; however, it has a very long reach, and on a large scale, it is the dominant force that shapes our Universe. The first three forces are unified, but scientists are finding difficulties in unifying all four. In particle physics, it is theorized that each of the four forces is carried by a specific particle.

The two main physics theories used to study our Universe today are quantum physics and general relativity. It is believed that the unification of these two fields holds the answer to many questions. The standard model of particle physics uses quantum theory and special relativity (but not general relativity) and unifies three of the four forces. In 2012, the Higgs boson, responsible for giving particles mass, was discovered at CERN. To study gravity, astronomers look at the sky. In contrast, particle physicists look inside the atom for a tiny particle that is believed to carry the gravitational force, the graviton (which is undiscovered yet). Both methods still need to yield a complete understanding of the four forces and a unification theory that can describe the Universe on both a large and a small scale. There are some promising theories out there, such as String Theory, as well as other alternative theories to general relativity that are being studied and could describe the behavior of the Universe on a large scale. However, none of these alternative theories are confirmed yet, and many of them fail to describe what general relativity describes today.

⋮

Future of the Universe

After talking about the beginning, the evolution, the composition, and the models of our Universe, we will now talk about its future. There are several possibilities determining the future of the Universe and its ultimate fate. The Big Freeze, the Big Rip, and the Big Crunch are the most common scenarios.

In the Big Freeze scenario, the expansion of the Universe continues, and stars continue to form for the next 1 to 100 trillion years, but eventually, the gas supply that is needed to form the stars will be depleted. The existing stars will also run out of fuel and ultimately die. Black holes will dominate the Universe and eventually evaporate through the Hawking radiation process. The Universe will continue to expand asymptotically until it finally grows into a dark, cold Universe and approaches absolute zero temperatures.

The Big Rip scenario is similar to the Big Freeze, but it is more extreme. In this scenario, the expansion of the Universe continues to increase until all objects are ripped apart, down to the particle level. The current expansion rate increases space between galaxies, but it is insufficient to rip them apart.

In the Big Rip scenario, the expansion rate will tend toward infinity, and everything will eventually disintegrate into elementary particles. In the Big Crunch scenario, it is believed that there is enough gravity in the Universe for objects to start coming closer to each other instead of being pulled apart. The expansion will decelerate and stop, and the Universe will collapse into what it was at the Big Bang, a singularity. Following this scenario, another Big Bang might occur. Big Bangs, expansions, and Big Crunches may eventually happen repeatedly. Here, the Universe would possibly be in an oscillatory state; it will keep on expanding and collapsing again and again.

These scenarios are the most common, however, several models can be derived from each scenario. Current data on the expansion and currently known physics suggest that the first two scenarios, the Big Freeze and the Big Rip, are the most likely to happen. I would also like to mention that there are alternate and complementary theories in cosmology that state that our Universe is just one of countless others. There are several ways these Universes can coexist. For example, they can occupy the same space but exist in different dimensions, or each Universe can be its own bubble that collapses and expands in the Multiverse. As there is little evidence to support the remainder of the theories, I will stop now.

Chapter 9

Cosmic Fireworks

I dedicate this section to talking about some phenomena that are the most extreme, explosive, and violent events in the Universe. These phenomena are often seen and studied in the high-energy domain and are often the products of compact objects. In astronomy, they are referred to as transients. Transients, as their name indicates, are one-time or recurring events that can vary from the nearby passing of a comet to a very far cosmic explosion. In the galactic and extragalactic fields, when we usually talk about high-energy transients, they mostly fit in the cosmic event category. Such cosmic events cannot be replicated on Earth and can be considered natural science labs for physics exploration. As you've seen in this book, many extraordinary things can be found in space. Here are a few more.

Flaring Stars

A flaring star occurrence is similar to solar flares in our system. It is the dramatic increase in the brightness of a star which can be seen in several electromagnetic wavelengths. The flares occur when the magnetic energy accumulated in the star's atmosphere is suddenly released. These flares are radiation bursts that can affect the planets and worlds around a star. On Earth, a solar flare may damage satellites and communication devices. In more extreme cases, flares could ionize a planet's atmosphere affecting its weather, habitability, and lifeforms. Flares could weaken a planet's magnetic field and strip away its atmosphere over time.

Novae

As explained in Chapter 4, most stars are not individual stars but are in binary systems. To have a nova event, two objects are needed: a star and a white dwarf in a binary system. In this scenario, at the origin, there were two stars, until one of them died a few billion years later and became a white dwarf. The star and the white dwarf are orbiting each other, and when they get close, the white dwarf accretes matter from the star due to its strong gravity. This matter falls into the white dwarf and forms a temporary atmosphere as illustrated in Figure

9.1. The white dwarf heats this atmosphere until it reaches a critical temperature. It eventually ignites, explodes, and is violently expelled into the surroundings. This type of event is named nova; Latin for "new star". This is because it was so bright that in the past that when astronomers witnessed these events, they thought that a new star suddenly appeared. Novae occurrences are common and can happen periodically in the same system following the orbital period of the two objects. Galactic novae, which are nova occurring in our galaxy, happen approximately 10 times per year. These events are usually detected by telescopes, but sometimes (around once a year), they are bright and close enough to be visible to the naked eye.

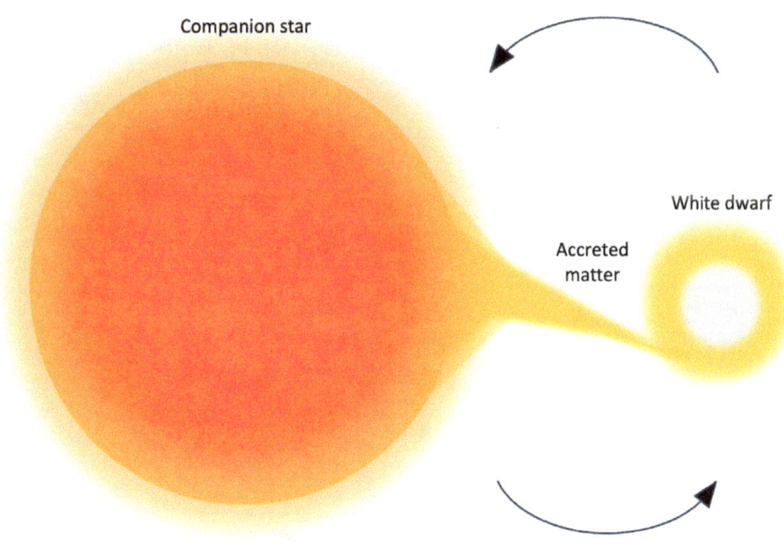

Figure 9.1

:
•

X-ray Binaries

X-ray binaries form when a star is coupled with a neutron star or a black hole. Matter falls from the star (the donor) to the neutron star or the black hole (the accretor), releasing significant amounts of energy in the form of X-rays detectable by X-ray telescopes. Within this category, microquasars are a subclass of X-ray binaries. The accretor in microquasars forms a collimated jet of high-energy particles with emissions ranging from radio (in particular), to gamma rays. Microquasars are named like that because they have similar features as quasars, but are smaller. Microquasars are on the scale of stars, while quasars are on the scale of a galaxy. Black holes are responsible for microquasar jets, while supermassive black holes are responsible for quasar jets.

:
•

Supernovae Type 1

Type 1 supernovae occur in a star-white dwarf system. The white dwarf is the product of the end of a Sun-like mass star's life. As explained in Chapter 4, the electron-repulsive force prevents a white dwarf from collapsing under the influence of gravity. However, when the white dwarf has a star companion, it can accrete its matter, increasing its mass. When it reaches 1.44 times the mass of the Sun, it becomes so massive that the electron-repulsive force is no longer enough to prevent it from gravitational collapse. The white dwarf loses its equilibrium and collapses in a one-time destructive event known as a type 1 supernova. The name "supernovae" comes from these events being much brighter than regular novae.

Type 1 supernovae are one of the most luminous events in the Universe. All type 1 supernovae events have the same brightness since they all happen at the same moment: when the 1.44 solar mass limit is reached. This means that the same amount of matter, 1.44 solar masses, explodes whenever there is a type 1 supernova. Due to this, type 1 supernovae are often used in astrophysics as cosmological probes. Since we know that all type 1 supernovae have the same absolute brightness (the brightness of an object as if it were at a distance of 10 parsecs), by measuring the amount of light (the apparent brightness) we receive from one and comparing it to the amount of light we would receive it at 10 parsecs, we can determine how far away the supernovae is and perform cosmological measurements. This also allows us to probe the distribution of matter in the distant Universe. One should note that type 1 supernovae are rarer than regular novae. They can happen only once in the lifetime of a white dwarf and are destructive to the host object. It is estimated that a type 1 supernova occurs once every several hundred years in our galaxy.

I want to remind you that even if an event is rare in a galaxy, it still happens quite often in the Universe because there are billions and billions of galaxies. The Universe has countless stars that can create these occurrences. Supernovae are one of the most powerful events in the Universe. When one occurs, the supernova's light typically outshines the total light coming from the galaxy. The emission rises over several days, then drops and remains for several years.

Supernovae Type 2, 1b and 1c

Type 2, 1b, or 1c supernovae occur when a massive star undergoes core collapse, as described in Chapter 4. When a massive star reaches the end of its life, it explodes and shatters its matter across billions of kilometers. The energy released in this event is equivalent to the total energy that a star can produce during its lifetime. You might be wondering why this terminology was chosen for supernovae. It would be more logical to choose 2b and 2c instead of 1b and 1c. However, these names are chosen because when analyzing the light coming from these supernovae, type 2 supernovae had hydrogen lines indicating the presence of hydrogen in the exploding star. In contrast, types 1a, 1b, and 1c have no hydrogen lines in the spectra. Type 1a supernova does not have hydrogen lines because, understandably, the exploding white dwarf does not contain any. Even though the core collapse of massive stars causes type 1b and 1c, they don't have hydrogen lines because, in these particular cases, the star was stripped of its outer hydrogen envelope before collapsing (due to rapid rotation, for example).

Core collapse supernovae events are more common than type 1 supernovae. Remember what it takes to have a type 1 supernova: It requires a binary system, and a low-mass star (that has a relatively long lifespan) that has already died and become a white dwarf; which takes a few hundred million to billions of years. During this period, the white dwarf accretes matter from the other star, which can take several million years.

Type 2 supernovae occur when a massive star (with a relatively short lifespan) dies, which only takes, more or less, a few millions of years. The last observation of a nearby supernova occurred in 1987. A massive star exploded in the Large Magellanic Cloud, our neighboring dwarf satellite galaxy. It was the first supernova that contemporaneous astronomers could study in real-time and in detail. This supernova released colossal amounts of energy in the form of electromagnetic radiation that was seen from Earth. However, it also generated colossal quantities of neutrinos. The Kamiokande detector in Japan detected 11 neutrinos from this event. Before this event, neutrinos from extra-solar origin were never observed. This was the first confirmed detection of extra-solar neutrinos. This was the beginning of neutrino astronomy.

Neutrino astronomy is the science of studying space through neutrinos. Neutrinos are particles with shallow masses that rarely interact with anything in the Universe and can escape the densest places, where even light is trapped. Neutrinos hold clues about the physical processes that created them. In the case of the 1987 supernova, they indeed confirmed the stellar evolution theory that described the end of life of a star. Supernovae leave behind supernovae remnants, the remains of the stars composed of gas and dust, shattered around in the form of nebulae. One of the most famous supernovae was observed in 1054; it was a galactic supernova 6,524 ly away. It was so bright (and close) that it could be seen by the naked eye, during the day, for weeks. The explosion's light was detected approximately 1,000 years ago from 6,524 ly away, meaning that this event occurred around 7,500 years ago, and the light took 6,524 years to reach us. This explosion left behind the Crab Nebula. It is a little less than a thousand years old and is still expanding under the effects of the blast. It is one of the most studied astronomical objects, and it is a reference to astronomers in the high-energy astrophysics field. Nebulae, like the Crab, can continue to expand, dissipate, or collapse under gravity's effects and form new stars with recycled gas and dust.

The last galactic supernova we know of is 28,000 ly away, on the other side of the galaxy. The light from this supernova should have reached Earth around a century ago, but it was hidden behind the galactic plane. However, the supernova remnant was discovered later on and was named G1.9+0.3. Technically this was not the last supernova that happened in our galaxy. Remember that this is 28,000 ly away, meaning it took the light 28,000 years to reach us. It happened 28,000 years ago, way before the Crab Nebula supernova, but the light only reached us around the beginning of the 20th century.

Compact Object Mergers - Gravitational Wave Events

Compact object mergers detected today can be the merger of two black holes, two neutron stars, or a black hole and a neutron star merger. These events are detected today on Earth through gravitational waves and are called gravitational wave events. As explained in Chapter 1, gravitational waves are generated by the orbital movement of heavy bodies. So what is better than massive black holes and neutron stars to generate this kind of emission? In a system including two compact objects like black holes or neutron stars, the two objects rotate around each other. These systems can come from old binary stars where both heavy stars collapsed and gave a neutron star or a black hole at the end of their life. In another scenario, one of the compact objects can capture another one under the effects of gravity. In any way, the two compact objects deform space-time around them and come closer and closer together. The whole system loses energy and radiates ripples (gravitational waves) in space-time. Over several million years, the two objects get closer and closer. A few seconds before merging, the two objects become so close that their orbital periods become in the order of milliseconds. Imagine two objects several times heavier than our Sun rotating that fast. During these final seconds, a substantial amount of gravitational waves is released, and the two objects eventually merge.

In the case of two black holes, the result of the merger is a black hole. Even though this is one of the most powerful and energetic events in the entire Universe, black hole mergers are silent. Black holes are dark and do not emit any light. The same is believed for their mergers. However, some theories predict electromagnetic radiation in extreme scenarios where the black holes are submerged in interstellar gas, for example. Do not get me wrong, when I say these events are silent, they still release gigantic amounts of energy in the form of gravitational waves. They are powerful enough to create ripples in space-time that can be detected on Earth with one of the most sophisticated and advanced technologies. The mass converted into energy during this process is comparable to our Sun's mass and can be equivalent to several times the mass of our Sun during the coalescence. These kinds of gravitational waves are detected on Earth through gravitational waves interferometers. The detection techniques are described in Chapter 11. The first detection ever made was in 2015 by the Laser Interferometer Gravitational-Wave Observatory (LIGO) collaboration.

Black hole mergers are not only typical of stellar black holes; two supermassive black holes in the center of galaxies can also merge. This happens when two galaxies merge. The instruments to detect these mergers are being commissioned and expected to make their first detections in the next decade. In the case of supermassive black hole mergers, the amount of energy released in the form of gravitational waves is at least a thousand times bigger, depending on the mass of the supermassive black holes. The energy release is spread through millions of years and can be detected for a long time with the planned instruments. It is believed that intermediate-mass black holes exist between 100 and 1,000 solar masses. However, it is unclear how such black holes would form as stellar physics does not allow the production of stellar black holes above 100 solar mass. Moreover, no observations were made until 2019. In 2019, the merger of an 85 solar-mass black hole, with a 65 solar-mass black hole was detected. The resulting black hole weighed 142 solar masses (with the rest radiated as gravitational waves). This was the first time the birth of an intermediate-mass black hole was observed. This proved the existence of such objects and provided hints on how they could be produced in the Universe: through mergers.

In the case of mergers involving neutron stars, the situation for the merger and gravitational wave release is similar. It is, however, different in the sense that electromagnetic emissions are expected this time, and the light from the explosion can be detected from Earth in the right conditions. The mass of neutron stars is smaller than black holes, so the energy released is smaller. However, neutron stars, unlike black holes, emit light. When two neutron-stars merge, the output can be a black hole if the neutron stars are massive enough. If not, then the new object created is a massive neutron star. It is useful to mention here that the techniques of gravitational wave detection do not allow us to have a good localization of the event in the sky. Think of it this way: instead of locating them, we are listening to them. When you hear thunder, you can roughly say where it came from. You can maybe indicate the direction, but you cannot say exactly where it came from unless you see the lightning. This is similar to when we detect gravitational wave events. We can say in which region in the sky the merger happened, but we cannot exactly pinpoint the location where it happened. We will have to scan large regions in the sky to search for their electromagnetic emission.

Only two neutron star mergers have been confirmed, and electromagnetic waves were detected only from the first one, in August 2017. This was the first detection of neutron star mergers, and it was close enough (130 million light-years away!). The event is called GW170817, and it opened a new era in multi-messenger astrophysics. Multi-messenger astrophysics uses more than one messenger to study space, in this case, gravitational waves and electromagnetic waves. A couple of seconds after the gravitational waves were detected on Earth, a burst of gamma rays (called gamma-ray burst) was detected by our satellites (Fermi and INTEGRAL), and 11 hours later, the optical light was detected. The radio and X-ray emissions arrived a few days late and lasted for years.

The information from all these messengers permitted us to study the science behind this event in ways that were impossible before. It had many implications in stellar astrophysics as it allowed the study of stellar evolution in unprecedented ways. In nuclear physics, it allowed the study of the structure of ultra-dense matter and the synthesis of heavy elements. For theoretical physicists, it provided unparalleled tests of general relativity. It was also a remarkable event for high-energy astrophysicists, with the study of high-energy emissions coming from the outflow of the explosion and the jet of high-energy particles created by the merger product (more details later). Moreover, detecting electromagnetic counterparts to GWs is extremely interesting for cosmology. For example, matching the estimation of the distance in two different ways using the information carried by the gravitational waves and the information carried by the light helped us understand the expansion of the Universe in new ways.

GW170817 is not the first gravitational event that included the detection of two neutron stars, but it is the first one detected with interferometry. In the 1970s, two astronomers indirectly observed gravitational waves by observing two pulsars, the Hulse-Taylor binary. They noticed that the distance between the pulsars was getting smaller and smaller over time, and deduced that the system must be losing energy through gravitational wave radiation.

Gravitational waves open new possibilities to study space as they can help us study and understand invisible events, like the merger of black holes. I want to also mention that other kinds of events can generate gravitational waves. For example, the merger of white dwarfs, the rapid rotation of pulsars with bumps on their surfaces, and supernovae. For now, the technology we have only allows the detection of gravitational waves generated by black holes or neutron star mergers, but the situation will surely evolve.

Kilonovae

Kilonovae occur when neutron-stars merge. It is one of the features of these merger events that we discussed earlier. Right after the merger, the heavy elements in the neutron are heavily bombarded by neutrons from the neutron stars. These elements undergo radioactive decay, leading to electromagnetic radiation emission in all directions. The signature radiation will rise in the first few hours, and then fall. The first hints of kilonova emission were discovered in 2008, then in 2013. However, the first confirmed kilonova came in 2017, accompanying the first gravitational wave event involving neutron star mergers, GW170817. The dual detection of the merger in both gravitational and electromagnetic waves confirmed the nature of the electromagnetic emission as coming from a kilonova and solved a long-lasting mystery. Kilonovae are sometimes accompanied by another transient electromagnetic event: short gamma-ray bursts.

Gamma-ray Bursts

In the 1960s, during the Cold War, the United States Air Force sent the Vela satellites to space to monitor any gamma-ray emission from nuclear explosions on Earth. These satellites were equipped with gamma-ray and X-ray sensors, and their goals were to monitor compliance with the 1963 Partial Test Ban Treaty by the Soviet Union. On July 2nd, 1967, the detectors caught a gamma-ray signal, but this gamma-ray detection did not have the distinct features expected from a manufactured nuclear event. It was clear that this event was not of terrestrial origin. The other possibilities were solar flares or supernovae, however, there were no detections of such events at the time of the gamma-ray signal. The U.S. army sent more Vela satellites which led to the discovery of 16 similar gamma-ray signals in total. All this information was classified at the time, and it wasn't until 1973 that the classification status was lifted, and the discovery of these gamma-ray bursts (GRB) was made public. These signals were from unknown cosmic origins, and the progenitor source was still indefinite.

Several missions were sent later on to study the origin of gamma-ray bursts. A debate arose in the science community on the origin of these events. Some believed they were produced inside our galaxy, while others thought they were from extragalactic origin. In the 1990s, the BATSE instrument on board the Compton Gamma Ray Observatory showed that the distribution of the gamma-ray bursts in the sky was isotropic; they were evenly distributed spatially. If these events were from a galactic origin, we would see a concentration toward the galactic plane, where most galaxy stars are. This is the first proof that gamma-ray bursts have an extragalactic origin. Later on, scientific advances permitted the localization of some gamma-ray bursts to their host galaxies, which was proof of their extragalactic origin.

There are two types of gamma-ray bursts: long gamma-ray bursts and short gamma-ray bursts. Long gamma-ray bursts have an initial emission phase that lasts several seconds, but their afterglow can last for months and years. They are believed to come from the core collapse of massive stars and are associated with supernovae. When a star collapses, in the first moments when a black hole forms, the latter accretes matter around it, and launches a collimated jet of gamma rays and charged particles. The jet is beamed, and the particles inside the jet travel at near-speed-of-light velocities. These jets are highly energetic. They were detected initially in gamma rays, hence the terminology. In fact, they can emit electromagnetic radiation, from radio to gamma rays. Nowadays, these jets are seen mainly in X-rays and low-energy gamma rays.

The second type is short gamma-ray bursts. The physical phenomena behind these jets are similar to the long ones, except that these bursts originate from the merger of neutron stars. The initial emission phase lasts less than 2 seconds, with an extended afterglow phase. Short-gamma ray bursts can also be accompanied by kilonovae emission, coming from the same event (merger of neutron stars), but produced by a different physical process (radioactive emission). Gamma-ray bursts, in general, are very energetic events in the Universe, but short gamma-ray bursts are usually less energetic than long gamma-ray bursts. The luminosity of a gamma-ray burst can reach 100 billion, billion times that of the Sun.

I want to note that new progenitor sources of gamma-ray bursts, like magnetar outbursts, are discovered today. Gamma-ray bursts, kilonovae, and compact object mergers are rare phenomena. They are expected to occur less than once in a hundred years within our galaxy. All the detections made to date of these three types of events are extragalactic. The most energetic gamma-ray burst was detected recently on October 9th, 2022, GRB221009A. It is nearly an order of magnitude more luminous than the second one on the list (GRB130427A) of gamma-ray bursts occurring until that date. The event was named the BOAT (Brightest Of All Time) by the media. If a gamma-ray burst is close enough to Earth, and the jet is pointed toward us, there would be a significant chance of wiping out life on the planet.

On a final note, some long gamma-ray bursts are discovered to have the same features as short gamma-ray bursts and vice versa. This begs the question: Is the classification gap between long and short gamma-ray bursts getting narrower?

Soft Gamma-ray Repeaters

Soft Gamma-ray repeaters are objects, most probably magnetars or other highly magnetized neutron stars, that display an outburst of gamma-ray emission from time to time. These outbursts are called flares, and sometimes giant flares. Flares from soft gamma-ray repeaters are very energetic. The word soft only comes from the fact that the gamma rays emitted belong to a low-energy gamma-ray range in the electromagnetic spectrum. Giant flares can be extreme, to the point that one of them, in 2004, perturbed the upper parts of Earth's atmosphere. Flares are triggered by a shift in the magnetic field of the magnetar, causing the crust of the highly dense dead star to break and lead to an outburst of X-rays and gamma rays.

Fast Radio Bursts

As their name indicates, Fast Radio Bursts are very bright flashes of radio waves that last from a fraction of a millisecond, to a few milliseconds. These very bright flashes are confirmed to be of astrophysical origin. However, the exact sources of these flashes are still unknown to this day. The first discovery of a fast radio burst occurred in 2007 by two astronomers looking at archival data from 2001, from the Parkes radio telescope. This first burst was named the

Lorimer burst. The emission peak was so bright that astronomers had difficulties believing it was astrophysical and suspected a terrestrial or artificial origin. However, the astronomical origin was confirmed.

After 2007, Astronomers started detecting these mysterious bursts regularly. Several communities of scientists formed around the globe to solve the mystery of the source of fast radio bursts. Several radio telescopes dedicate large portions of their observation time to detect and follow these events. Telescopes observing all electromagnetic ranges also follow up on fast radio bursts. To this day, the riddle of fast radio burst is still unsolved. We know that there are two types of fast radio bursts: the first ones are non-repeater fast radio bursts that occur only once, while the others are repeating bursts from the same source. It could be that all of them are repeaters with delays of many years, and we still haven't detected their repetitions.

Until 2018, the number of theories and models describing the possible sources and origins of fast radio bursts exceeded the number of fast radio burst detections. These models range from associating them with known physical phenomena like magnetar outbursts and mergers of neutron stars, to more far-fetched phenomena, including alien spaceships, exotic models, and cosmic strings. Since 2018, the number of fast radio burst detections has increased drastically with the installation of a new radio telescope in Canada called CHIME. This telescope alone is responsible for detecting more than 500 confirmed fast radio bursts to date. Around 20 fast radio bursts are associated with host galaxies, but it is still unknown where the rest have come from.

I should mention here that it is also challenging to know precisely the distance of fast radio bursts; we can, however, estimate their minimum distance.

In 2020, a fast radio burst, FRB 20200428A, was detected for the first time from within our galaxy. It was associated, for the first time, with a soft gamma-ray repeater, SGR1935+2154; a magnetar. Moreover, for the first time, X-ray outbursts from the magnetar were associated with the detected radio emission. This dual detection positioned magnetars as primary progenitor candidates for fast radio bursts.

Tidal Disruption Events

A Tidal Disruption event happens when a star orbiting a supermassive black hole gets ripped apart as it gets closer to the black hole. In this scenario, a portion of the star is accreted by the black hole which can form an accretion disk, flares of electromagnetic emissions, and possibly neutrinos. The star is not always destroyed entirely and can survive the supermassive black hole encounter.

AGN Flares

Active Galactic Nuclei (AGN) flares occur when these objects undergo a high activity period. This usually happens when a big chunk of matter is accreted by a supermassive black hole that launches a powerful jet of relativistic particles. The phenomenological reasons behind this massive accretion can vary. For example, the passage of a gas cloud near the black hole, and its accretion under the effect of the gravitational pull of the black hole, can be one of the reasons. Other reasons can be due to gravitational interactions between galaxies pushing big chunks of gas into the supermassive black hole.

AGN flares are very luminous. A typical flare is several billion times more luminous than the Sun. Of course, the fact that they are very far away explains why we don't see them shining in the sky without the use of a telescope. AGN flares can feature neutrino emissions. Gamma rays can be produced by physical phenomena involving the acceleration of leptons (electrons) to very high speeds near the speed of light. On the other hand, gamma-ray production can be produced by heavier particles than electrons, called hadrons (like protons). The theory shows that neutrino emissions can accompany these hadronic emissions.

In 2017, a gamma-ray flare coincident with neutrino emissions from a blazar called TXS 0506+056 was discovered for the first time. This hinted that hadronic interactions could be involved in the AGN processes generating gamma rays. This is an example of how multi-messenger astronomy can help us understand the physics of the objects around us. By combining information from the two different messengers, neutrinos and electromagnetic gamma-rays, our understanding of the gamma-ray emission processes and particle interactions in AGN jets improved drastically. This dual detection also helped confirm the astrophysical origin of the neutrinos detected, which is very difficult and rare to prove in general.

224

Chapter 10

Humankind in Space

Astronomers usually study celestial objects using observations from ground-based or space-based observatories. For nearby objects, like planets and asteroids in our solar system, we can send space probes to fly by or orbit them, and robots to land on their surfaces to study them. Human exploration is even possible for very close objects like the Moon (and potentially Mars). Until now, the farthest a human has ever gone is the Moon. The International Space Station orbiting Earth is constantly occupied by astronauts visiting, with specific missions that usually last a few months. In the following chapters, I introduce the most prominent space programs, missions, and instruments that we, humans, have created to pursue space. In this chapter, I talk specifically about the adventures of humankind in space.

First Firsts

The USSR was the first to successfully launch a satellite, Sputnik 1, into orbit in 1957. After that, came the challenge of sending humans to space. During the Cold War, the USA and USSR were in a race to send the first man to space. Several animals were sent first, such as mice, frogs, guinea pigs, monkeys, and dogs into the depths of space. The most famous animal to be sent to space by the USSR is Laika, the dog, but unfortunately, Laika didn't make it back. Eventually, the USSR succeeded in sending a man to space first. On April 12, 1961, they sent cosmonaut Yuri Gagarin, the first man to journey in low Earth orbit. 23 days after Yuri Gagarin, USA sent their first man to space, Alan Shepard, on May 5th, 1961. In the same month, John Kennedy, the president of the USA, announced that his country would land a man on the Moon before the end of the decade. The first woman in space was Soviet cosmonaut Valentina Tereshkova in 1963. Note that there is no difference between an astronaut and a cosmonaut except that the latter is trained by the Russian space agency while, the first is trained by the United States, Canadian, European, or Japanese space agencies. A Taikonaut is trained by the Chinese space agency.

Unfortunately, ever since Sputnik 1, humanity has left a lot of trash in space. Earth's surroundings are now filled with millions of debris pieces composed of leftovers of decommissioned or failed satellites. These remains orbit Earth at colossal speed reaching several thousands of kilometers per hour. A small nail in space, hitting a satellite or a space station at that speed, could cause significant damage. Space debris is now monitored and avoided in space missions. However, a more sustainable solution is needed. Some agencies deorbit their satellites after their service is finished to limit this ecological catastrophe.

.
.
●

The Moon Landing

The Moon landing came after nearly a decade of preparations. The project that eventually took humanity to the Moon for the first time was the Apollo project. The Mercury project preceded Apollo and it was the project that sent the first American man to space; the Mercury capsule can only fit one person, while the Apollo capsule was designed for three people. Gemini, an intermediate program serving two crew members, ran between 1961 and 1966. The Apollo project ran from 1961 to 1972, but unfortunately not without fatalities. In 1967, the Apollo 1 cabin caught fire on the ground during a rehearsal test, killing three crew members. The first successful Moon landing came with Apollo 11, in 1969.

Apollo 11 mission used the Saturn 5 rocket, which was one of the largest rockets ever built. The Apollo 11 mission, the Saturn 5 rocket, the astronaut suits, the lunar module, and all other aspects of this endeavor are engineering wonders that deserve a comprehensive standalone book to do justice to their sophistication. The Apollo 11 mission was launched on July 16, 1969; it reached the Moon after a three-day journey. The Apollo capsule had two compartments, a command Module and a Lunar Module. When the mission arrived on the Moon, the command Module stayed in orbit around the Moon with pilot Michael Collins on board. The Lunar module detached and headed to the surface with Neil Armstrong and Buzz Aldrin, landing approximately 6 hours later. Neil Armstrong was the first man to set foot on the Moon on July 21, 1969, at 02:56 GMT. He is the first human to "walk" on the surface of another celestial body. In his words: "That's one small step for [a] man, one giant leap for mankind". Buzz Aldrin followed him shortly after. The two astronauts stayed 21 hours and 36 minutes on the surface of the Moon. They collected 21.5 kg of Moon material and then rejoined the command module. Launching the lunar module from the surface of the Moon to join the command module in orbit, was one of the riskiest parts of the mission. If it failed, the two astronauts would have no way of returning home. After a successful rendezvous, the three astronauts returned to Earth and landed on July 24th, 44 hours after leaving lunar orbit. Five crewed missions followed Apollo 11. Between 1969 and 1972, six crewed missions were sent to the Moon, from Apollo 11 to Apollo 17. After an oxygen tank failure, the lunar landing was aborted during the Apollo 13 mission. The three astronauts could not land on the Moon but returned safely to Earth. The last crewed mission to be sent was Apollo 17 on December 7, 1972. After that, Apollo 18 and 19 were canceled due to budget cuts; humanity never returned to the Moon since then.

The total lunar landing efforts cost at the time was nearly 26 billion dollars, equivalent to more than 250 billion dollars today (adjusted for inflation). Going to the Moon is more complicated than one would think. The first men to go to space were orbiting a few hundred kilometers from the surface. The Moon is more than 380,000 km away, more than 1,000 times farther. It is like the difference between going to the supermarket and traveling to another country. NASA and its worldwide partners (like the European Space Agency, ESA) plan to return to the Moon. They are planning to send the first woman on board another one of the largest rockets ever built, the Space Launch System (SLS). The mission is named Artemis, and the first crew test flight is scheduled for no earlier than 2024. In 2022, Artemis 1 was launched to the Moon in an uncrewed test mission.

Space Stations

The Moon is not the only target for humans. Humankind also ventured to space, where they established a home. Today thousands of artificial satellites are orbiting Earth. Some are pointed toward Earth and used for weather, military, positioning, communication, and other purposes. Others, like space telescopes and space detectors, are pointed toward outer space to study it.

Other artificial satellites are built to be crewed; I am talking about space stations. In 1971, the USSR built the first space station, Salyut 1, which stayed in orbit for around 6 months. It orbited 200 km from the surface and resembled a 20x4 meter cylinder. Two crewed missions visited the space station with a total of six cosmonauts. After the crew of the second trip was asphyxiated during re-entry, the program was aborted, and the space station was deorbited; it was purposely lowered into the atmosphere to burn and disintegrate.

Several space stations followed Salyut 1: Salyut 3 to 7, by the USSR, and Skylab by the USA. The last space station sent by the USSR was Mir, in 1986, which was later operated by Russia. Mir was assembled in space and is the first space station composed of several modules. It was also the first space station continuously inhabited, and lasted for a long time. Mir was, at the time, the largest artificial satellite built by humanity. It had a capacity for three crew members, and it orbited between 300 to 420 km above the surface and completed 15.6 orbits per day. It was then decommissioned and deorbited in 1996.

International Space Station (ISS) later succeeded the Mir space station. It is a joint project between the USA, Canada, Japan, Russia, and Europe. The first component was sent to space in 1998, and the first long-term occupants later arrived at the ISS in the year 2000. ISS is the longest-term inhabited human-made satellite and it is still active, with a record of 23 years today. The ISS is by far bigger than any previous artificial satellites in terms of size. It is composed of 17 modules (the last one was installed in 2021) and has a length of 109 meters and a total mass of 445 tons. It is also one of the most expensive projects built by humankind, with a cost of 150 billion dollars. China also entered the space station race. Since 2011, they have sent two small space stations Tiangong 1 and 2, both decommissioned in 2018 and 2019, respectively. In 2021, they launched the Tiangong space station, which has been in orbit for more than 2 years, and has the capacity for three crew members.

So, what are space stations used for? From a practical point of view, the simplest answer is that space stations are microgravity labs. Crew members of the space station conduct experiments and gather data in all fields, like biology, human biology, physics, astronomy, engineering, and meteorology. ISS houses five detectors of high-energy particles and radiation to gather astrophysical data. The research goals can vary from understanding how life is in microgravity, to developing new technologies useful for permanent space colonization.

Space Vehicles

We have discussed humanity's most distant trips, space stations, and satellites orbiting our planet, but we have not yet discussed the vehicles that send satellites, materials, and humans into space. To send something into space, it is usually stored in a capsule and launched into space on top of a rocket. Rockets are, for now, our primary vehicles to go to space. They can be fascinating when they are not used to transport bombs. I cannot talk about all the rockets, but I can briefly overview some of the ones I like. We can divide rockets and space crafts into two categories: those used to transport humans, and those used to transport material and satellites.

A lot of major countries have their own space programs and rockets used to send material into space. From those, I mention Ariane, the European (French origin) rockets that can be considered at the top of the list of reliable and safe rockets. Ariane 5, the one currently in use, sends the heaviest cargo to space. If space rockets are constructed with utmost safety measures, then manned space rockets must be even safer. Only a handful of rockets that can take astronauts into space were made.

The Saturn 5 rocket, used to send humanity to the Moon, is one of them. I will only mention a few other spacecraft classes that marked human travels to space. The first one is the Russian Soyuz, the successor of Vostok, that took the first man to space. Soyuz refers to both the rocket and the capsule. The Soyuz rocket is the launch vehicle that transports the Soyuz capsule, which carries astronauts and cosmonauts into space. Soyuz is currently the longest-standing spacecraft used to send humans into space, and it is still operational today. The Soyuz rocket's first launch was in 1986, with the Soyuz capsule fitting three crew members on top. Often the Soyuz spacecraft can be seen docked on the ISS as it is one of the primary vehicles to transport both astronauts and cargo (food, for example).

Another spacecraft is The Space Shuttle is a USA-made engineering marvel. The vehicle resembles an airplane that is mounted on three rockets. After the launch, the rockets fall back to the surface using a parachute, and the shuttle continues its journey to space. After its re-entry, the shuttle lands on the surface like an airplane and uses parachutes to decelerate. The Space Shuttle brought flexibility and maneuverability to human spaceflights. Not only was it used to send the Hubble space telescope, but it was also used to service this telescope 5 times with crewed missions, which is impossible to achieve with space capsules. After two fatal crashes, the space shuttle program was interrupted in 2011 due to cost and feasibility reasons. The first disaster occurred in 1986 when the Challenger space shuttle exploded a few minutes after launching and killed all seven crew members. The second disaster occurred in 2003 when the Columbia space shuttle disintegrated upon re-entry, also killing all seven crew members.

After the space shuttle, Soyuz was the only remaining vehicle to send humans to space for nearly 10 years. This was before the interest of private industries in space began.

In the second millennium, space became interesting for the private sector. Several companies started their attempt to build and send their space crafts to space. Some of them are meant for space tourism. Space X is one of the most developed companies that produce space vehicles. They revolutionized the industry by building reusable rockets that can safely land back on Earth. NASA uses Space X's Falcon 9 rocket to send cargo and astronauts to space today. The Falcon Heavy is designed to carry heavy loads. Finally, the Starship spacecraft, which is mounted on the Super Heavy rocket, has the ultimate goal of sending humans to Mars sometime in the 2030s.

Space Probes - Voyagers

Humanity sent space probes to all the planets in the solar system, and to other objects like asteroids, comets, and dwarf planets to study them. The ingenuity in these missions, such as the landing maneuvers of the rovers on Mars and the operations and science preparation behind the Cassini-Huygens mission to Saturn, require an entire book to be well described. However, I promised myself to keep this book short. Therefore, I will only mention two space probes that I find to be the most interesting and the most intriguing:

Voyager 1 and 2. As for the rest, some are mentioned in Chapter 3.

Voyagers 1 and 2 were sent to look closely at the outer planets and their moons. Voyager 1 was launched on September 5, 1977, with the Titan-Centaur rocket. Voyager 2 was meant to travel further and was launched first, on August 20th of the same year.

Five main instruments are used to collect the scientific data on the Voyagers to study the magnetic field, the plasma, the cosmic rays, and the electromagnetic emissions from the outer planets and interstellar space. Voyager 1 reached Jupiter in March 1979, after a two-year-long journey. Voyager 2 arrived in July of the same year. Voyager 1 and Voyager 2 reached Saturn in 1980 and 1981, respectively. The two probes discovered new rings and new moons of the two planets. After Saturn, the two probes parted ways. Voyager 2 still had two more worlds to visit: Uranus and Neptune. It discovered and photographed many of the planet's rings and moons and discovered the dark patch on Neptune.

The plan for Voyager 1 was to send it as far as possible toward the edges of the solar system. In 1981, Carl Sagan had the idea to ask NASA to turn a Voyager spacecraft around and take a last picture of the Earth. It was difficult at the time to do so; the image had no scientific value, and there was a risk that the instruments would be damaged by the radiation of the Sun. However, Sagan's proposition was eventually put in motion, and in 1990, the picture was taken from a distance of 6 billion kilometers. In the image, a pale blue dot appears; this is Earth. Carl Sagan said: "Consider again that dot. That's here. That's home. That's us." Voyager 1 continued its journey to the edge of the solar system. It crossed the solar system's boundaries and entered interstellar space in 2012. Traveling at more than 61,000 km/h, Voyager 1 was the farthest-ever human-made object from Earth. Voyager 2 crossed into interstellar space in November 2018.

The two missions were first designed to last 5 years. More than 45 years later, the probes are still functional and sending valuable data on the interstellar medium to Earth. Both probes are in power-saving mode now. Eventually, their energy supplies will end, and we will not be able to communicate with them anymore. Voyager 1 is around 24 billion km away now, and Voyager 2 is about 18 billion km away. Even after their power runs off, they will continue their path in the depths of space. They will still be traveling even after we are all dead. The probes are not headed to the nearest star, but they might encounter some systems on their way. In 40,000 years, Voyager 1 will pass 1.6 ly away from the AC+79 3888 star in the constellation of Camelopardalis. Around the same time, Voyager 2 will pass at 1.7 ly near the Ross 248 star. In 300,000 years, it will pass 4.3 ly away from the famous Sirius star. Both probes carry a copy of a golden record where a message from humanity is recorded in case they encounter any intelligent extraterrestrial life. These records are intended to promote the diversity of cultures on Earth; they include greetings in 55 languages, famous music from known artists, and pictures and sounds of people and places on Earth.

All the crewed and uncrewed missions described in this section were meant to study objects in the solar system and its surroundings. However, humanity is aiming way beyond the solar system. A project, the Breakthrough Starshot, aims to send tiny probes to the closest system, Proxima Centauri to take pictures of the system, collect data and send it back to Earth. Proxima Centauri is 4.25 ly away. If the Voyager probes, which are one of the fastest manufactured objects ever built, take thousands of years to reach nearby star systems, how are they going to make it happen in a human lifetime? Well, the idea of the project, founded by Yuri Milner, Stephen Hawking, and Mark Zuckerberg in 2016, is to send light sails probes that will be powered by lasers from the ground. A light sail is like a wind sail, except that the energy from light acts as the pushing force. The lasers will accelerate these tiny probes to a quarter of the speed of light. This will allow the Starships to complete the journey to Proxima Centauri in around 20 years. When the probes send the data to Earth, the signal containing the data will travel at the speed of light. This means the data will take 4.25 years to reach us after it is sent. The project is still in the commissioning phase and is expected, if successful, to launch in 20 years. Given that the timeline will be respected and adding the 20-year journey and 4.25 years for the data to reach Earth, we will get the first images in around 40 years. Of course, scientists and engineers on this project will face many challenges, such as decelerating the probes when they reach Proxima Centauri if needed to take the images.

Chapter 11

Space instruments

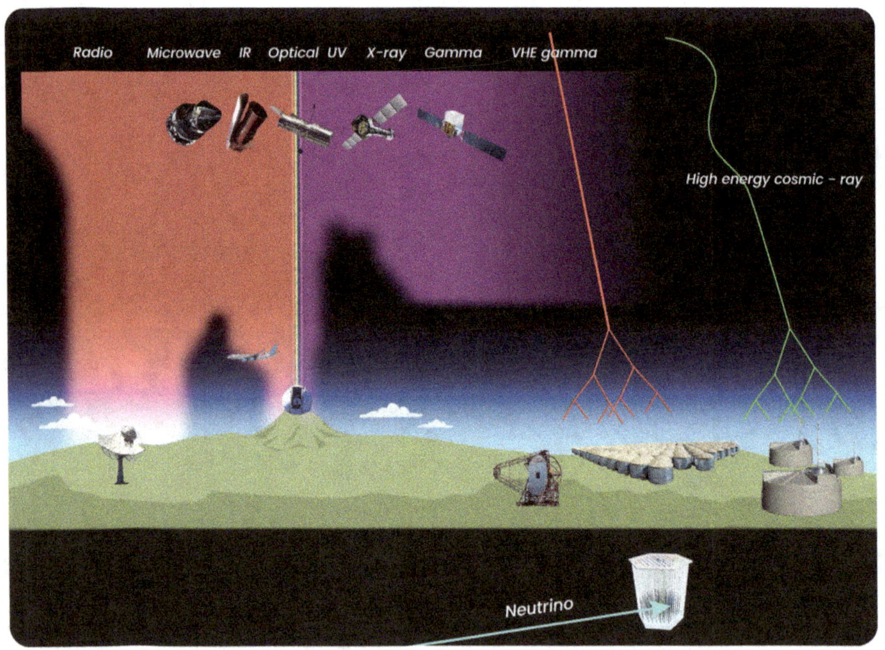
Radio
Microwave
IR
Optical
UV
X-ray
Gamma
VHE gamma
High energy cosmic - ray
Neutrino

Humanity's quest to explore space has led to numerous breakthroughs in our understanding of our solar system. However, with today's technology, it is extremely difficult to go beyond our neighborhood. Moreover, sending space missions is exceptionally costly and takes a very long time. Despite this, scientists and engineers have gained valuable insights beyond the solar system, and far into the mysteries of the Universe through observatories on Earth and around it. In the following sections, I give an overview of how humankind uses different messengers from space to study the Universe.

.
.
●

Observations

The first messenger that we receive from space is electromagnetic radiation. When we talk about electromagnetic emissions, we can think of light. Electromagnetic radiation is visible light, in addition to the radiation in different energy bands that we, as humans, cannot see. From the lowest to the highest energy, the electromagnetic bands are radio, infrared, optical, ultraviolet, X-rays, and gamma rays. From these, only the optical light is visible to humans, while the others require special instruments to detect them. There is a duality around the nature of electromagnetic radiation. Since the 1800s, following the works of James Maxwell and, later on, Heinrich Hertz light has been considered a form of electromagnetic emission that propagates in waves with different wavelengths. Electromagnetic emissions are characterized according to their wavelength and energy. To understand the concept of wavelength, think about it as the width of a wave in the ocean. Similarly, electromagnetic waves can be characterized by their wavelengths. The smaller the wavelength, the more an electromagnetic wave can penetrate objects (between the atoms), hence, the greater the energy.

This was until de Broglie, a French physicist, discovered that light also can behave as a particle. In this scenario, to understand how light can act as a particle, I like to imagine that the light coming out from a light bulb is in the form of billions and billions of small pellets bouncing everywhere in the room through the darkness, and eventually entering my eyes. So, is light a wave, or a particle? Apparently, it is both. Today we talk about the dual nature of light. This may not be easy to understand, but it is the basis of quantum mechanics, and believe me, quantum mechanics is not easy to understand if we reference our high school physics logic.

For the curious reader, a famous example of quantum mechanics is Schrodinger's cat. Imagine you have a cat in a box. You don't know if the cat is alive or dead. In the logic of quantum mechanics, the cat is both alive and dead until you open the box and see for yourself. It is the same for light; it is both a wave and a particle until we take a measurement to determine its nature. Depending on our measurement, light will exhibit itself either as a wave or as a particle.

For the upcoming sections, what is essential to know is that electromagnetic radiation comes in different sizes (wavelengths), and that the electromagnetic spectrum is divided into seven ranges. The bigger the size of the electromagnetic wave, the smaller its energy. From the biggest to the smallest size:

- Radio waves, often used for communication.
- Microwaves, which are often considered as an extension of radio waves, are used in electrical appliances and communication.
- Infrared, which can be used to measure the heat of our bodies.
- Optical, which contains the seven colors that we can see.

- Ultraviolet, which burns our skin when we are in the Sun for a long time.
- X-rays, used in medicine to study the interior of the body, have small wavelengths and can penetrate our skin.
- Gamma rays, which are emitted in nuclear experiments.

Celestial objects can emit electromagnetic radiation in all of these electromagnetic ranges, and they can also cast other messengers, but let's stick to electromagnetic radiation for now. Each type of electromagnetic radiation gives different information on the object we are studying. For example, radio waves can be used to trace the existence of Hydrogen, whereas infrared light can be used to trace the presence of heated dust. To study the physical processes taking place in space, there are several methods we can rely on. I have previously mentioned some of these methods, but I will restate them as a reminder.

Imagery

We can rely on imageries to study the morphology of objects. If we take an image of a galaxy using optical light with an optical telescope, the regions that emit light will appear in our image. The principle behind cameras in astronomy, often charged coupled devices (CCDs) for optical telescopes, is that they count the number of photons inside each pixel, and build a spatial image of the observed object. A higher number of photons in a given region is translated into a higher intensity in the camera. This is how images are taken.

I want to mention that professional and advanced cameras used in astronomy do not show colors. As black and white images, the results are only of high and low-intensity regions. The color is then artificially added using filters. By using a blue filter, only the areas emitting blue light will be shown in the black-and-white image. By using another filter — green, for example—the regions emitting green light will be shown. The blue and green colors are then both added to the images, which combines the final shot. The regions that emit light in a galaxy are usually full of stars; therefore, imaging a galaxy will show us the stars in this galaxy. However, there are also regions of dust clouds where the dust blocks the starlight behind it and restricts us from seeing anything. However, the dust is heated by the high temperature of the stars, and like any other object, the dust emits strong infrared emissions when heated. Unfortunately, even by using an optical telescope, these infrared emissions are not visible to us, but we can still use a dedicated infrared telescope to detect them. Therefore, combining the images taken with both an optical and an infrared telescope can give us a more comprehensive picture of the galaxy. A similar approach can be followed if we want to image the surface of a planet or any other object.

Spectroscopy

Another way of studying astrophysical phenomena is by calculating spectra. A spectrum shows the composition of the electromagnetic radiation coming from a source. To do so, we use spectroscopy, where the different components of electromagnetic radiation are measured.

A prism decomposes the light into seven colors. Measuring a source's light spectrum shows us the light's composition (refer to figure 4.5), and the different colors coming from that source. Moreover, if some elements exist between the source and the observers on Earth, some parts of the light will be absorbed or enhanced. For example, suppose we are studying a planet with oxygen on its surface. In this case, the oxygen will absorb some parts of the blue light heading toward us. We will find that in our spectrum, there are some missing parts in the blue light emissions. This indicates that oxygen exists on this planet.

A spectrum of sources can be created across all wavebands. For example, suppose we want to study the physical processes behind an AGN flare. In that case, we can compute the light energy distribution of the emissions from this AGN using different instruments in different wavebands. We will find that the AGN has very high X-ray emissions. The X-rays can be generated either from a heating process or from the fact that there is a strong magnetic field where charged particles are accelerated. By looking at the shape of the X-ray spectrum, we can differentiate between the two.

A heating process and a particle acceleration process have different theoretical X-ray spectra. We can determine which fits the data better, by comparing our measurements with the theoretical expectations of how the spectrum should look. Suppose our measured spectrum resembles the theoretical heating process spectrum. In this case, it is most probable that the sheer energy of the object is heating the material inside the AGN. Now, suppose the measured spectrum resembles the predictions of a spectrum by the particle acceleration process. In this case, it means there is a strong magnetic field in the vicinity of the AGN, and there is a process accelerating particles to high energies.

We can combine these measurements with the measurements of gamma rays. Very high-energy gamma rays are difficult to produce, and should primarily be produced by acceleration mechanisms. If we detect very high-energy gamma rays from this AGN, it confirms the existence of particle acceleration in a strong magnetic field. Looking closely into the X-ray and gamma-ray measurements will also help us determine how strong the magnetic field is, how many particles are accelerated, and how much energy the AGN releases.

Photometry

Photometry measures the amount of light coming from a source in a given electromagnetic band. Several properties of celestial objects can be deduced by performing simple photometry. For example, there is a relation between the mass of the star and the amount of light it emits. By measuring the amount of light, we can deduce the mass of a star.

Photometry can be used to study physical phenomena by computing light curves. This means that we look at the variation in the amount of emission from a source as a function of time (refer to figure 4.6).

This measurement can be done across several periods and can indicate how the source and the physical processes behind it are evolving. For example, supernovae are bright in the beginning but fade with time. Let's look at a supernova to know how old it is. By measuring the amount of light, and how it varies with time, and then comparing it with theoretical models, we can know the age of the supernova, and when it occurred. We can also compare the apparent brightness of the supernova with its absolute brightness, which allows us to estimate the distance of the supernova. Another excellent example of how light curves are used in astronomy is the detection of exoplanets. Suppose we are looking at optical light curves from a star and we notice a slight dimming in the light of this star, this could mean that there is an object passing between us and the star that is occulting the light; possibly a planet.

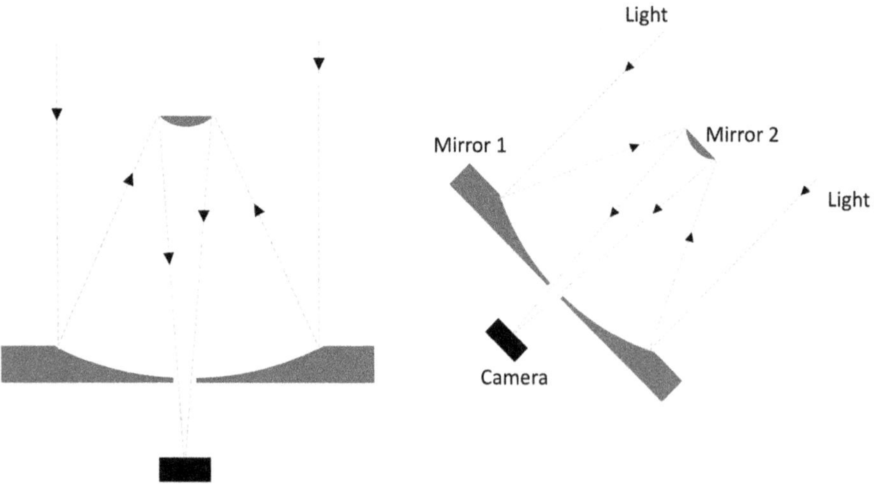

Figure 11.1

Let's now dig into the world of instrumentation in astronomy, and explore the machines that allow us to take these kinds of measurements. Let's venture into the world of telescopes. First of all, what is the concept of a telescope? A telescope is a device used to collect and read a signal. It is used to collect, and see light from the cosmos. The bigger the collection area of a telescope is, the more signal it can collect. To understand it better, we refer to the example previously given and to Figure 11.1. Light particles can be thought of as rain droplets, and telescopes are like buckets. If we want to collect more rain, we need larger buckets. The larger the collecting area of a bucket is, the more it can accumulate rain droplets. If a telescope can accumulate more signals, fainter objects can be revealed. Your eye retina is typically less than 1 cm in diameter. The largest optical on Earth has collecting areas of around 10 m in diameter. This means that these telescopes have a collecting surface that is more than a million times larger than the human eye, and more than a million times more powerful. Nowadays, light-collecting areas of telescopes are well-smoothed mirrors. The light is reflected by the mirrors toward a reading device.

Radio Astronomy

Electromagnetic signals in the radio wavebands have large wavelengths that can vary from a few millimeters to several hundred meters. They are large enough to be reflected on metal surfaces and do not require mirrors; this is why radio antennas are just a piece of metal.

There are several types of radio telescopes, but the most typical one is the radio reflector. It consists of a large dish of metal that collects radio signals from outer space and reflects them toward a receiver device. The receiver then converts the signal into an electrical signal that is read by a detector and converted into an image. In a particular region of the sky, the brighter the region is, the higher the electrical signal will be, and the brighter the part of the sky will be in the image. Since a metal dish is relatively inexpensive, radio telescopes can be constructed in big sizes. The bigger the dish is, the bigger the collecting area is, and the more it can pick up faint signals. In addition, the bigger the radio dish, the better its resolution.

Nowadays, tens of radio observatories are observing the skies around the globe. The largest one is the FAST telescope in China, with a dish diameter of 500 m. It is preceded by the 300 m Arecibo telescope in Puerto Ricco, which has now been decommissioned. Radio telescopes can also work together as arrays where the power of several telescopes can be combined to mimic a larger telescope. For example, some interferometry techniques allow combining the power of two telescopes, 100 m apart, into almost one equivalent 100 m-sized telescope. The most extensive array of radio telescopes that is being built for the future, is the Square Kilometer Array (SKA), deployed on two sites: in Australia and South Africa. Predecessors of SKA are MeerKAT in South Africa and ASKAP in Australia. These two arrays are composed of several telescopes and are one of today's most powerful radio observatories. MeerKAT, for example, is composed of 64 dishes. Not all radio telescopes have dishes; some of them are just composed of small antennas. The principle is the same: the antennas collect the signal, but the signal will be combined and read differently this time. In this case, the time difference in the signal's arrival between the antennas plays a significant role in the data recombination. These antennas are fixed to the ground, and have the advantage of surveying large sky areas at once. The data analysis is then focused on a particular region. In contrast to antennas, dishes are focused on a specific region in the sky. Of course, radio dishes have other advantages, like accurately computing the direction of a signal. SKA is an effort of more than 100 organizations in over 20 countries. The final array will comprise thousands of dishes and antennas spread over a square kilometer.

Other notable radio observatories are the 64 m Parkes telescope in Australia, which discovered the first Fast Radio Burst, and the Nançay large radio telescope in France. Nançay radio telescope (NRT) has a primary reflector of a 200x40 m long rectangle shape reflector that sends radio signals toward a secondary 300x35 m parabolic surface. Other radio telescopes are ATCA in Australia— composed of six 22 m dishes—the 100 m Green Banks radio telescope, and CHIME, the Canadian observatory, which is currently responsible for detecting the large majority of fast radio bursts. Today, an international effort is put into combining several telescopes and spreading them across the globe to simulate an Earth-sized radio telescope. This telescope is called the Event Horizon Telescope (EHT).

EHT was able, for the first time in the history of humanity, to take a picture of the supermassive black hole in the center of the M87 galaxy, back in 2019. The trick here was to gain enough resolution power to see the black hole in a galaxy far, far away. Even though the supermassive black hole is colossal, it is still very difficult to see since it is very far. By building this Earth-size telescope, astronomers were finally able to do it. Remember, enlarging the collecting area gives a better resolution, and allows the collection of more signals. In 2022, EHT was able to image the supermassive black hole in the center of our galaxy. It is only the second direct image of a black hole ever produced.

A final note on radio astronomy is that, unlike visible light, radio waves can be detected during the day and are not occulted by sunlight. This means that radio telescopes have the advantage of being operable both day and night and have a very high-duty cycle.

Optical Astronomy

In this section, I discuss optical telescopes. Infrared, optical, and ultraviolet detection techniques are more or less the same; therefore, I will combine their discussion in this section.

To observe visible light, there should first be darkness; otherwise, the sunlight will occult all the faint light coming from celestial objects. This is quite obvious since we cannot see stars during the day. In the past, optical telescopes resembled large monoculars that were composed of a lens that concentrated light toward an eyepiece device. These telescopes are called refractors since the lens refracts the light. As mentioned before, the larger the telescope is, the more powerful it can be. It is very difficult to build large lenses, but there is a way around it; we can use mirrors instead. Telescopes that use mirrors are called reflectors. The idea is that light is reflected by a mirror toward a reading device. The reading device is either our own eyes or a camera. Nowadays, cameras are used in astronomy to have long exposures, allowing light accumulation.

The construction of the mirrors constrains the size of optical telescopes. The mirrors used in optical astronomy are very high-end mirrors with polished surfaces at the nanometer scale, which is 50,000 times smaller than the thickness of a human hair. These mirrors are very costly, and the additional bending effects should be considered the bigger the telescope is.

The largest optical telescopes found today are the Keck telescopes in Hawaii with 10 m mirrors, and the Grand Telescopio Canaria (GTC) in La Palma with a 10.4 m mirror. These mirrors are composed of small segmented mirrors. Segmented mirrors have a big advantage, as they are relatively easier to produce. Still, their effectiveness is smaller than full mirrors. Other notable telescopes are the Very Large Telescopes (VLTs) operated by the European Southern Observatory (ESO) in Chile. The VLTs are four large 8.2 m telescopes and four small 1.2 m telescopes that can be operated together. Using interferometry, the VLT is one of the most powerful observatories on the ground. They operate in infrared and optical wavebands.

Optical telescopes should be operated in total darkness. Therefore, sites with low light pollution are favored. Several sites on Earth meet these requirements. Some of the most telescope-populated areas are the summit of Mauna Kea on the Big Island in Hawaii, the Atacama Desert in Chile, and the Roque de Los Muchachos Observatory in La Palma.

Two new projects are being prepared to build the largest optical telescopes on the ground: The Extremely Large Telescope (ELT), a 40 m European telescope currently under construction in Chile, and the Thirty Meter Telescope (TMT), an American project that will consist of building a 30 m telescope in an undetermined location.

Optical telescopes are largely constrained by the atmosphere, so the disturbances in the atmosphere largely affect the data and do not permit a good resolution. Therefore, some technologies, like adaptive optics are applied for large telescopes to correct the effect caused by the atmosphere. Adaptive optics change the shape of a reflecting mirror to adapt to the variations caused by the atmosphere. However, there is another way of getting rid of the atmospheric effects by just getting rid of the atmosphere as a whole and going to space. Sending an optical telescope to space has advantages, as this telescope will be operable both day and night, and the atmospheric effects will be non-existent. In addition, these telescopes operate in ultraviolet wavebands, which is very difficult to achieve on Earth since ultraviolet radiation cannot easily penetrate our atmosphere. But can you imagine the cost and the difficulty of sending a 10 m large telescope to space? It is practically undoable, but relatively small-sized telescopes, on the other hand, can be built and sent to space.

One of the most powerful telescopes we have today is the Hubble Space Telescope (HST). It is a 2.4 m diameter telescope that supersedes the capabilities of the largest telescopes on the ground. HST operates in infrared, optical, and ultraviolet wavebands. The Hubble telescope was sent to space in 1990, and it was initially intended to be operational for 15 years. The first images of Hubble showed that the telescope had a defect, and its first images were aberrated (distorted). The Hubble telescope was serviced and fixed thanks to the space shuttle program. After that, the telescope underwent four servicing missions, making it the most powerful telescope 30 years after its launch. One of the biggest discoveries of HST is the Hubble Deep Field. This image was taken when a group of scientists suggested pointing it toward an empty region in space. This region is the size of a tennis ball seen from a 100 m distance. After weeks of exposure, HST discovered thousands of galaxies lying in this region, showing that the Universe is much larger and richer than we previously thought.

As am I was writing this book, the successor of Hubble was being commissioned. The James Web Space Telescope (JWST) is a 6.5 m telescope sent to space at the end of 2021. Once scientific operations began, it became by far the most powerful optical telescope built by humankind. JWST is operated mainly in the infrared domain which makes it the successor of Hubble and Spitzer; another noticeable space infrared instrument. Infrared light can escape dusty regions where optical light is trapped. JWST will be able to observe galaxies at the very edge of our Universe. The light of the farthest galaxies is so redshifted that the bulk of the emission is no longer in the optical domain but in infrared. These galaxies would then be invisible to an optical telescope, no matter how large it is, but not to JWST which sees in the infrared. The total cost of the project is somewhere around 10 billion dollars, and it is one of the most anticipated astronomy projects. It is an effort by NASA, ESA (European Space Agency), and CSA (Canadian Space Agency). The mission accumulated a delay of 10 years, partially because the telescope had to be sent to a point in space -1.5 million km away from Earth. At this point, and with the decommissioning of the space shuttle program, nothing can be done to fix it if something goes wrong. Therefore, the construction of the telescope and the launch must both be made with perfection. JWST's first image was revealed in July 2022 by the president of the USA. This image is a deep field image, like the ones that Hubble produced, but deeper. In a few hours, JWST did what Hubble would need weeks to do.

Finally, I would like to mention that there are optical observatories intentionally built to conduct large-scale surveys. Here are a few examples. A 2.5 m telescope conducts the SDSS survey in Mexico, which has computed images for nearly 50 million galaxies so far. The Euclid mission is a space-based telescope that is set to launch in the year 2023 and is estimated to survey several billion galaxies with a high impact on dark energy studies. The ZTF facility uses a very large camera that is mounted on a 1.2 m telescope to detect transients in the night sky. More than 6,200 confirmed transients have been detected by ZTF so far. The Vera Rubin Observatory, formerly known as LSST, is an 8.4 m telescope equipped with the largest camera that is being built in Chile. This telescope will take transient surveying (and dark energy studies) to another level. The Kepler space telescope is a space-based observatory that detects and surveys exoplanets by the transit method. To date, the Kepler telescope discovered more than 2,600 exoplanets (more than half of our inventory). Gaia is a space-based instrument that surveys the stars of the Milky Way to build the largest catalog of stars and map the galaxy. Unlike other telescopes, the Gaia observatory has no mirrors and directly captures the star with its large cameras onboard.

High Energy Astronomy

X-rays and gamma-rays are energetic electromagnetic emissions, with high frequencies and energies, but have small enough wavelengths to easily penetrate the human body. X-ray and gamma-ray telescopes are bound to be exclusively space-based because they cannot penetrate our atmosphere. While this is good for humanity, X-rays and gamma rays are still harmful to the human body; this complicates the task for X-ray and gamma-ray astronomers who must send their telescopes and observatories to space. For low-energy X-rays, telescopes operate similarly to optical telescopes, but with different mirrors and cameras. For example, X-rays are very energetic and they pass through materials. The mirrors of the X-ray observatories are arranged in canonical layers, nearly parallel to incoming X-rays, looking like glass barrels. This way, the incoming X-rays skip on the mirrors toward the detector. Having several layers separated by a few millimeters allows to enlarge the collecting area. Today, NASA's Chandra and EU's XMM Newton telescope are the most famous X-ray telescopes in space. These, accompanied by a large arsenal of X-ray telescopes, permit us to discover the skies in the X-ray bands. Data from these telescopes are widely used in high-energy astrophysics. Studying higher energies requires the implementation of different techniques. The amount of high-energy X-rays and gamma-rays produced by astrophysical objects are usually smaller than in other bands (since they require rare high-energy astrophysical phenomena), but the study can be made on a one-by-one basis to achieve the required results.

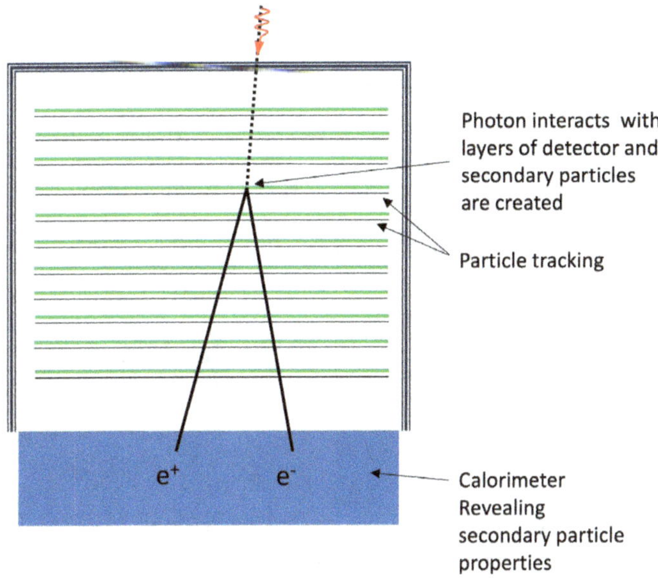

Figure 11.2

Space observatories that are meant for these observations do not usually have mirrors because the photons can pass through them. Their detectors are made from materials that interact with high-energy photons (different materials for different energy bands) as shown in Figure 11.2. The product of these interactions allows us to deduce the properties of the initial photon, like the direction and energy. This is similar to optical astronomy but with larger detectors more focused on single-photon detections, with different technologies and different data analysis techniques.

The Fermi observatory (Illustrated in Figure 11.2) is currently the most important gamma-ray observatory in space. Fermi is also used to detect gamma-ray bursts and is one of the most active instruments today. Swift is another telescope dedicated primarily to the detection of gamma-ray bursts. Swift has a monitoring instrument with a large field of view on board, and two telescopes in the X-ray and ultraviolet bands to take additional precise measurements once the monitor detects a gamma-ray burst. Several other space-based observatories monitor the skies nowadays, like the European Integral, the Italian Agile, the Chinese HXMT, the Indian Astrostat, NASA's NuSTAR, and many others.

Several high-energy observatories are planned to launch into space in the upcoming decade. One of the largest X-ray missions is Athena, a European mission expected to launch in the 2030s. As for GRBs, a large-scale mission is still in consideration; however, for now, the successor of Swift is a small-sized French-Chinese mission called SVOM, which will be launched in 2023 with space and ground sections. These are some of the missions currently leading the high-energy astronomy field.

Very high-energy astronomy deals with very high-energy gamma rays. These are the most energetic electromagnetic radiations that are at the very far end of the electromagnetic spectrum. Detecting very high-energy gamma rays by current space observatories is not easily achievable. Fermi can pick up a few photons from the lower part of the very high-energy band. However, there is a very efficient indirect method of detecting those on the ground. Although high-energy instruments are majorly space-based, very high-energy instruments, illustrated in Figure 11.3, are mostly on the ground.

As previously mentioned, gamma rays are not able to penetrate the atmosphere. However, when gamma rays enter the atmosphere, they interact with it. Bi-product particles of these interactions will form, giving birth to air showers. The speed of light is slightly lower in water and air than in the vacuum of space. The shower particles, traveling near the speed of light, decelerate in the atmosphere as the speed limit cannot be exceeded. As they do so, they emit swift flashes of faint ultraviolet and blue light, called Cherenkov radiation. This product of radiation created by very high energy gamma-rays can be detected by sophisticated instruments on Earth. These instruments are called Cherenkov telescopes, and they use the atmosphere as a calorimeter. By detecting these very brief flashes of light and analyzing their characteristics, the properties of the primary very high-energy gamma rays are deduced. The telescopes are similar to optical telescopes since only optical light is observed directly. However, the mirrors are not required to be polished to the precision of optical telescopes, which gives the advantage of constructing large telescopes with segmented mirrors. Moreover, the cameras, composed of photomultipliers (PMTs), can read much fainter and much faster signals than optical astronomy standard cameras, since the flashes of light are very brief at the nanosecond scale.

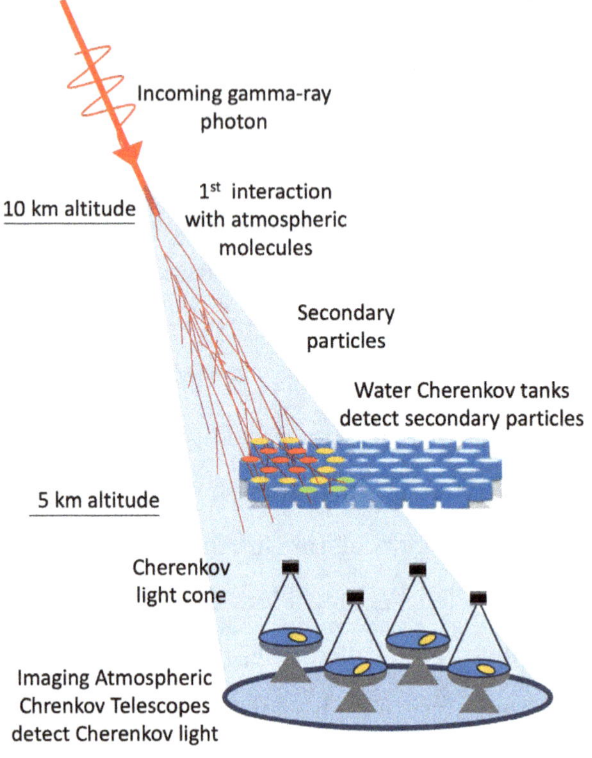

Figure 11.3

Today's largest Cherenkov experiment is the High Energy Stereoscopic System (H.E.S.S.). H.E.S.S. is comprised of four telescopes with 12 m segmented mirrors, and a fifth telescope with a 28 m mirror, which is the largest on the ground to date. The next generation of the Cherenkov Observatory is the Cherenkov Telescope Array (CTA) which will be built on two sites in La Palma and Chile. It will be an international observatory with 70 telescopes of different sizes in Chile, and around 20 telescopes in La Palma. The project started with the first large-scale telescope construction completed in 2021, in La Palma.

Another type of very high-energy observatories on the ground can be used to reach even higher energies, such as water Cherenkov tanks (Figure 11.3). These are water tanks that include light detectors. For the highest energy gamma rays, the air shower particle from the initial gamma ray will enter these Cherenkov tanks, where Cherenkov radiation will be emitted in the water with the same deceleration mechanism described above. The properties of the gamma rays are then deduced after a series of analyses, and the timing calculations are done between different tanks. The two largest water Cherenkov tank observatories are HAWC in Mexico and LASSO in China. Since both observatories are in the northern hemisphere, there is currently an international effort to build a water Cherenkov tank observatory (SWGO, for the Southern Wide-field Gamma-ray Observatory) in the southern hemisphere to monitor the galactic center that is only visible from the south. This is mainly because the galactic center houses the most energetic emission sources that can emit such high-energy gamma rays. Due to their detection techniques, Cherenkov telescopes must be built on high-altitude sites, around 2,000 m, and water Cherenkov tanks on sites with 5,000 m altitudes.

Cosmic-ray Astronomy

Cosmic rays are charged particles coming from outer space. These particles can be simple electrons and protons, or heavier atomic nuclei, like charged iron cores. Cosmic rays are the second category of messengers we detect on Earth after electromagnetic radiation. The amount of information they transport is significant to high-energy astrophysics. Nobel prize winner Victor Hess (whom the telescope was named after) was the first person to capture cosmic rays. Hess took his radiation measuring equipment, on board a balloon, and measured the radiation variation in Earth's atmosphere. These daring experiments led him to deduce that the radiation levels are greater at high altitudes. This meant that there were radiations penetrating our atmosphere from outer space. This radiation was named cosmic rays, by Robert Millikan in 1925. Since then, many discoveries in the cosmic ray field have been achieved, like the discovery of the positron, an antimatter particle (of the electron), and the muon, by Carl David Anderson, who shared the 1936 Nobel prize with Hess.

Cosmic rays extend a large band of energy. The less energetic ones have similar energies to gamma rays, and the highest energy ones can reach energies trillions of times greater than gamma rays. The heavier the cosmic ray, the more energetic it is. Cosmic rays have different origins. Some of them come from solar flares; those are the lowest energy cosmic rays. More energetic cosmic rays emerge from within our galaxy, but the highest energy cosmic rays are believed to be from extragalactic origin. The energetic cosmic rays are believed to be created by the most energetic phenomena in the Universe, and they are the highest energy particles that we know of today. Since cosmic rays are charged, they are affected by magnetic fields in space, and while traveling, they change direction according to the magnetic fields they encounter. In addition to the variety of magnetic fields that cosmic rays will encounter, our galaxy has a strong magnetic field which also makes it complicated to determine the direction of cosmic rays since their path has been affected by magnetic fields. This also makes it very difficult to determine the origin of the cosmic rays. Identifying the origin of the highest energy cosmic ray, called ultra-high-energy cosmic ray, is currently one of the most compelling dilemmas that high-energy and very high-energy astronomers are attempting to solve.

The detection of cosmic rays is achieved through instruments that measure the interaction of cosmic rays with other materials, which allows us to deduce the properties of the cosmic rays. Some of these instruments can be mounted on balloons and sent to the edges of the atmosphere. Some tools are sent to space, and some are attached to the ISS. For example, AMS-02 is an instrument that measures cosmic rays with a focus on antimatter particles, like the positron. Cosmic rays also generate Cherenkov air showers similar to gamma rays but in a more chaotic manner. A similar method to the water Cherenkov tanks is used to detect cosmic rays through their showers. Very-high-energy astronomers and cosmic-ray astronomers are often looking for similar things to try and answer similar questions. In fact, cosmic rays in the Universe produce a large amount of very high-energy gamma rays making it possible to either study the cosmic rays directly by detecting them on Earth, or indirectly by catching the gamma rays they emit at the source. Ironically, cosmic rays penetrating our atmosphere are a source of annoying background noise for very-high-energy astronomers, since they also produce air showers and Cherenkov light that can be confused with gamma-ray-induced showers.

The Pierre Auger Observatory in Argentina is the leading international experiment on ground-based cosmic ray astronomy, named after the person who discovered air showers from cosmic rays. The observatory is composed of 1,660 water Cherenkov tanks. The secondary particles, created in the atmosphere by interacting cosmic rays, enter these water tanks decelerating and emitting Cherenkov light. Since cosmic ray showers are more spread out than gamma-ray showers, the water tanks are also spread out in the field. The 1,600 water tanks are spread across a surface of 3,000 km², nearly one-third the size of Lebanon, at altitudes ranging from 1,330 m to 1,620 m. In comparison, HAWC comprises 300 water Cherenkov tanks across 0.022 km². Water Cherenkov tanks can be operated during the day but they will have low energy and spatial resolution. Cosmic air showers can also be detected by other means; for example, by observing a fluorescent light (different from the Cherenkov light) that cosmic air showers produce in the atmosphere. The same observatory is equipped with fluorescence and other air shower detectors that can only be operated at night. Pierre Auger is credited for discovering and studying the highest-energy cosmic ray particles. There are current plans to upgrade the Pierre Auger observatory into Auger Prime.

Neutrino Astronomy

Neutrinos, as their name indicates, are neutrally charged particles emitted by different physical processes. They were once thought to be massless, but they actually have shallow masses. Takaaki Kajita and Arthur B. McDonald shared the 2015 Nobel Prize for the discovery of neutrino oscillations, which shows that neutrinos have mass. Neutrinos travel at almost the speed of light. They are emitted in nuclear reactors and in the atmosphere but are also produced in astrophysical sources. The latter are the neutrinos that we will talk about here.

Neutrinos rarely interact with any form of matter; they only interact through the weak force and gravity. This is a double-edged sword for us. On the one hand, neutrinos can escape from dense places where electromagnetic radiation cannot escape, and they provide valuable information from these places. On the other hand, they are tough to detect because they rarely interact with any detecting material on Earth. In brief, neutrinos can penetrate anything and come out the other side. It is estimated that a lead-wall with a thickness of one light year is needed to stop half the incoming neutrinos. That is more than 2,000 times the distance between the Sun and Neptune. Since neutrinos can penetrate almost anything, they can come from very far away, and so most of the time, it is difficult to attribute their source.

Earth is bombarded by countless neutrinos every second. Trillions of neutrinos are penetrating your body right this second. Most of these neutrinos come from the Sun, some are human-made, and some come from far beyond. Astrophysical neutrinos, not generated by the Sun, are typically generated in energetic cosmic events such as supernovae or AGNs. The processes producing astrophysical neutrinos involve very high-energy cosmic rays. Neutrinos have an important value for astronomers. For example, scientists study the possible physical phenomena behind the high energy emission of AGN flares. There are two possibilities, either the gamma rays are emitted by accelerating electrons, or by accelerating more massive particles, like protons. In the case of the latter, the production of neutrinos is also expected. So, detecting gamma rays and neutrinos that are simultaneously coming from the same AGN, would prove the latter theory. This is one of the ways in multi-messenger astronomy, where information from different messengers is combined to study and understand the skies.

In addition, neutrinos can escape places where light is usually trapped. For example, in a supernova's first moments, the core medium is very dense as the star collapses, so light cannot escape; however, neutrinos can. It was the case in the 1987 supernova, where we detected a bunch of neutrinos a few hours before seeing the light of the supernovae in the Large Magellanic Cloud. The neutrino detection tells us the exact moment when the star's core will collapse. Light does not provide this information, since light emission is delayed because it cannot escape the dense core of the collapsing star in its first moments. Technically, the supernova was located in the Large Magellanic Cloud 160,000 ly away, meaning that the supernovae happened 160,000 years ago, but we just received the message a few years ago. The 1987 event is considered, by many, the beginning of neutrino astronomy.

Raymond Davis Jr. and Masatoshi Koshiba were awarded the 2002 Nobel Prize for the detection of cosmic neutrinos. They shared the Prize with Riccardo Giacconi, who was awarded the Prize for the discovery of X-ray cosmic sources.

Neutrino detectors nowadays are usually composed of several light detectors submerged in large masses of water or ice. The idea is that a neutrino, from the countless neutrinos traveling through the material (water or ice), will interact with it. This interaction creates secondary particles that emit Cherenkov light which the detectors will observe. The energy and direction of the neutrino can be determined by both the amount and the pattern of the light seen by several sensors.

IceCube is currently the largest neutrino observatory on Earth; it is designed to detect high-energy neutrinos. IceCube is comprised of 5,160 light sensors submerged in 1 km^3 of ice at the south pole. The sensors are deployed on lines below the ice, at a depth between 1,450 m and 2,450 m, and with 60 sensors for each line. Trillions of neutrinos from the northern hemisphere, penetrate the Earth from the north and travel through it until eventually, with some chance, a few of them interact with the ice on their way out, and get caught by the detectors.

Neutrinos from outer space usually have very high energies. The IceCube collaboration has claimed several neutrino candidate discoveries from astrophysical origin. A pair of neutrinos, nicknamed Bert and Ernie, were discovered in 2013. They are a thousand times more energetic than very high-energy gamma rays. In 2018, the IceCube collaboration announced the discovery of high-energy neutrinos from the TXS 0506+056 blazar in September 2017. This finding is simultaneous with gamma-ray discoveries from Fermi and MAGIC (a ground-based Cherenkov telescope array for very high-energy gamma-rays) collaborations. This discovery greatly impacted the study of physical phenomena generating gamma rays in astrophysical sources.

In 2022, the IceCube collaboration made a big announcement identifying neutrinos (accumulated through the years) from a galaxy called NGC 1068, which is 47 million light-years away, and in its center lies a cosmic accelerator of particles (probably a supermassive black hole). An upgrade that will take IceCube to the next level is ongoing, and it will take effect in 2026. By then, IceCube will be at least 8 times larger, and 5 times more sensitive.

Antares is a smaller neutrino observatory, submerged 2.5 km deep in the Mediterranean Sea. Its principle is the same as IceCube, but neutrino interactions occurred in seawater instead of ice this time. Antares is decommissioned now. It consisted of 900 optical modules divided among 12 underwater lines. The successor of Antares is

KM3NeT which consists of more than 10,000 optical modules and 600 lines and is being commissioned now.

From the other detectors, the Super-Kamiokande designed for lower-energy neutrinos, is one of the oldest large ones. It consists of a large tank of purified water, buried 1,000 m deep under a mountain in Japan. The tank's capacity is 50,000 tons of water, with 13,000 optical sensors inside to detect light from neutrino interactions. Kamiokande (its predecessor) started operating in 1983 and has undergone several upgrades since then. Kamiokande was not initially designed for neutrino astronomy but rather to study neutrinos on the particle physics side. However, the detector started being used for neutrino astronomy when it discovered the neutrinos from the 1987 supernova, where 11 neutrinos were detected coming from an astrophysical source for the first time. The successor of Super-Kamiokande will be Hyper-Kamiokande, a neutrino detector with the same traits, but a much larger size, and with a tank containing a million tons of purified water.

Gravitational Waves Astronomy

Gravitational waves astronomy is one of the newest fields of astronomy. Gravitational waves are ripples in space-time; the fabric of the Universe. Einstein first predicted their existence as a consequence of general relativity. Gravitational waves travel at the speed of light, and they are the fourth messenger we receive from space, after electromagnetic radiation, cosmic rays, and neutrinos. When a gravitational wave passes through, it stretches and compresses the fabric of space-time. Both space and time are compressed and stretched by the passage of a wave.

Gravitational waves are everywhere, but their effect is so small that we can only see the most energetic ones. The merging of compact objects, such as neutrons stars and black holes, produces these energetic gravitational waves. General relativity requires that the system be asymmetric; such that the masses of the two objects should be different to produce gravitational waves efficiently. The two objects initially orbiting each other will come closer together, and the system will release energy through gravitational wave radiation. Just seconds before they merge, the amount of energy released by gravitational radiation is colossal and enough to be detected by current technologies on Earth.

Other phenomena also produce gravitational waves in the Universe. They can be produced by rapidly rotating pulsars with defects (bumps) on their surfaces, or by cosmic phenomena, like the Big Bang, and many other things. At this time, we are being compressed and stretched continuously, but do not think of it like our bodies are being squeezed while the space around us remains the same. Instead, it is the space-time that our bodies occupy that is being compressed and stretched. This is the exact feature of gravitational waves that physicists use today to detect them. Imagine a road in front of you that is 1 km long: If a gravitational wave passes in the direction of the road, the road will be compressed for a brief time, and it will no longer be 1 km long during this time, but slightly less. A 1 km road will shrink by less than the width of an atom. This effect is very difficult to detect, which is why we do not notice gravitational waves when they pass by.

In the last 8 years, physicists finally caught gravitational waves successfully by continuously upgrading the detectors they have been working on for over two decades, and which they originally conceived more than five decades ago. Before we dive into the concept of gravitational waves detector, let me remind you that the first gravitational waves were seen indirectly, before the sensors came online. It happened in 1974 when two astronomers, Russell Hulse and Joseph Taylor, were observing a binary system where one of the objects was a pulsar. Pulsars emit signals in an exact timely manner, and by measuring the timing of this pulsar, they discovered that the orbital period between these two objects was getting smaller and smaller, meaning that the two objects were getting closer and closer. They deduced that the system must be losing energy by gravitational wave radiation, as predicted by Einstein nearly a century ago.

Nearly a century after Einstein predicted gravitational waves and half a century after the Russel-Taylor binary, physicists achieved a direct way of detecting gravitational waves. Gravitational wave detectors are interferometers. In interferometers, two beams of light are sent in different directions and then reflected by a mirror at the end, and then combined at the same point as shown in Figure 11.4. The interference between the two beams is observed. The wave nature of light depicts that if the beams are synchronized, the result will give bright light fringes, and if they are not synchronized, the results will give dark fringes.

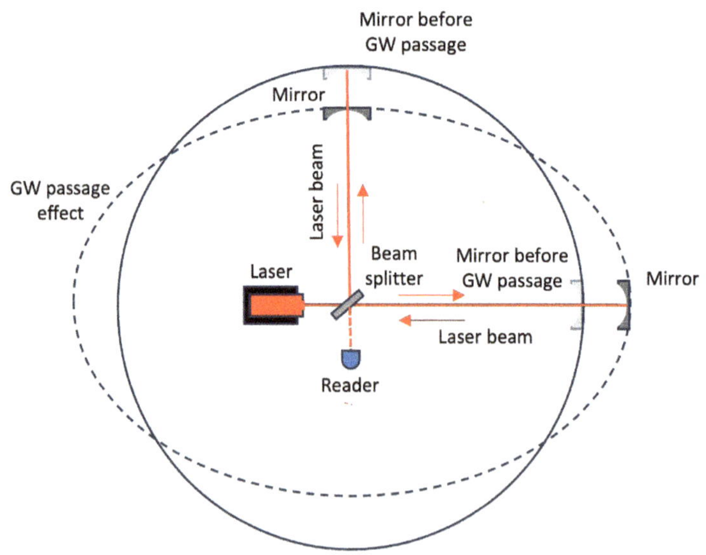

Figure 11.4

Gravitational wave detectors are giant complex interferometers. The first interferometer to come online is the product of the LIGO collaboration. LIGO collaboration has constructed two gravitational wave detectors in the United States; one in Hanford and one in Livingston. The two interferometers are composed of two giant structures, called arms, that are perpendicular to each other. Each arm is 4 km long. Two synchronized laser beams are sent from the same source, into the arms. The laser beams bounce on a mirror at the end of each arm, then return and combine again at the base of the arms. Suppose a gravitational wave is passing by in the direction of one of the arms. In that case, the latter temporarily shrinks, while the other one expands. This means that the laser beam in the first arm, travels slightly less than 4 km back and forth, while the other beam travels slightly more than 4 km back and forth. This causes the originally synchronized beams to become asynchronous, or vice versa, and the fringes formed by the two beams will change. This is how gravitational waves are detected.

The first gravitational waves were detected in 2015 by the LIGO interferometers. They were produced by the merging of two black holes 1.3 billion light years away; 1.3 billion years ago. During the detection, the path of the light was only changed by a width of an atom. To detect such small changes, gravitational wave interferometers are built with extreme care. Everything has to be extremely precise, and the sources of background noises have to be well-identified and isolated. The mirrors of these interferometers are the highest quality available. They are optimized to be affected by the least possible noise and vibrations and to reflect the most possible amount of light. They are the most stable manufactured objects.

Hundreds of scientists, in all fields, are joining forces to maintain and upgrade the interferometer from mirrors, laser beams, detectors, and software, to the overall structure. Gravitational wave interferometers are not operated continuously. To be able to upgrade and maintain them, they operate during a defined period: a few months to a year called an observing run. The first observing run was held from September 2015 to January 2016. This was the time during which LIGO detected the first gravitational waves. Rainer Weiss, Barry C. Barish and Kip S. Thorne were awarded the 2017 Nobel Prize for their contributions to the LIGO detector and the observation of gravitational waves.

In 2017, LIGO resumed science operations. Virgo is another collaboration with an interferometer in Italy. With 3 km long arms, Virgo is slightly smaller and less sensitive than the LIGO interferometers. In August 2017, Virgo joined LIGO in its second run. In August 2017, GW170817, the binary neutron star merger, was detected with its colossal implications on astrophysics, as discussed in Chapter 9. LIGO and Virgo collaborations joined forces and began observing a third gravitational wave run in 2019 and 2020. The third observing run lasted nearly a year but was cut short a few weeks before the expected end date, due to the COVID-19 pandemic outbreak. During the third observing run, LIGO and Virgo observed 80 gravitational wave events;

8 times more than in the previous runs, bringing the total detection up to 90 events.

KAGRA is another interferometer in Japan. It is the first underground gravitational wave interferometer with 3 km long arms. It was under commission during the third run and joined LIGO and Virgo toward its end. KAGRA will join the fourth run that started in mid-2023. The KAGRA sensitivity is currently negligible compared to LIGO and Virgo, but it will improve over time. KAGRA is planned to operate in cryogenic technology in the future, resulting in better isolation of the instrument from background noises. As I write, the astrophysical community, myself included, is preparing for this upcoming fourth run, where the detection rate is expected to increase even further. A third LIGO interferometer that is being built in India will come online around the end of this decade.

Two major projects are planned for the future (in the 2030s): The Einstein Telescope (ET) in Europe, and the Cosmic Explorer (CE) in the USA. ET will be an underground triangle-shaped interferometer with 10 km long arms. CE is still at an early stage in planning but is expected to consist of two L-shaped observatories: one 40 km in length and the other 20 km in length. These two future interferometers will take gravitational wave astronomy to the next level.

Gravitational waves we detect today come from the merger of stellar compact objects. Gravitational waves produced by the mergers of supermassive black holes are too large to be detected by current ground interferometers; the wavelength is just too large to be noticeable. One would have to build an interferometer with an arm-length of a million kilometers to detect such waves, which is obviously impossible to achieve on the ground. However, going to space is possible to help us achieve the unachievable.

As we speak, a space mission for gravitational waves is being prepared. LISA will consist of three space probes, equipped with laser and detectors positioned 2.5 million km apart. Together they form a triangle-shaped interferometer with a 2.5 million km arm length. The concept is the same as ground interferometers, with some differences in application. For example, LISA will not have any mirrors as it won't be possible to bounce a laser beam from that far away. Instead, each probe will have two lasers as well as light detectors on board. With an expected cost of 1.1 billion dollars (that will be reduced due to the current economic crises) and relentless design challenges ahead, LISA is planned to launch in 2037.

Chapter 12

Search for Extraterrestrial Life

Wow!

Is there life out there?

SEARCH FOR EXTRATERRESTRIAL LIFE

It may be that the most common question related to the Astronomy field is: Is there life out there? Many believe that seeking extraterrestrial life is the ultimate goal of many astronomy fields. However, if life exists outside Earth, it will not necessarily be similar to life on Earth. Even if it was similar, it would not necessarily be in human life, and if it was, then it would not necessarily have the same stages of evolution.

Extraterrestrial life, if it exists, can be anything; It could be microbial, plant, animal, or even advanced humanoid form, or it could even be none of the above. Life on Earth is carbon-based and requires specific elements for its survival. Life elsewhere could be different. It could be silicon-based, for example. In our current understanding, it could be so different that we could not recognize it as life. Our imagination here can take us anywhere. I could imagine giant planet-like organisms floating in space, multi-separate-body animals, purely energetic lifeforms, etc... However, for the rest of this section, let's focus on forms of life that might be similar to what we have on Earth: bacteria, viruses, plants, fungi, animals, humanoids, et cetera.

When we explore the solar system and the objects it contains, such as planets, moons, dwarf planets, and asteroids, we are systematically searching for life on these objects. One of the main goals of the many missions to Mars is to search for forms of life —mainly microbial — evidence indicating the existence of previous life, or essential elements for life, such as water and organic compounds. These searches are relevant, not only to Mars but to many other places in the solar system, like the Europa and Enceladus oceans discussed in Chapter 3. Although water has been found on Mars, and in several other places in the solar system, life or previous life forms have not yet been found. The exoplanet field is centered around the search for a home that could be suitable for life. Several candidates have been found, but we are yet to discover any proof of the existence of life in them. There are other means of searching for life, especially in its intelligent form.

SETI, short for Search for Extraterrestrial Intelligence, is a set of scientific searches for extraterrestrial intelligent life; signs of radio transmission from alien civilizations. SETI uses radio telescopes to search for radio signals that aliens may use to communicate with each other, or with us. SETI has used several radio telescopes in the past. Currently, SETI searches the skies for alien signals using some existing radio telescopes, like the giant FAST radio telescope (with SETI as a core objective), and the Green Banks telescope. SETI also has its dedicated radio telescope array, called the Allen Telescope Array (ATA), previously known as the One Hectare Telescope (1HT), with an array equivalent to a 100 m radio dish. SETI also targets other forms of alien communication indices. For example, there is a program for the search for optical light from lasers that might be used in alien communication, and another program for the search for signatures of extraterrestrial artifacts and technologies in the vicinity of the solar system and in extrasolar systems.

Although the many dead ends might come as a bit of a disappointment, there are, however, some rare and exciting moments. For example, in August 1997, a strong 72-second-long signal coming from the direction of the Sagittarius constellations was detected by the Big Ear radio telescope in the United States. While exploring the data, Jerry R. Ehman discovered bright signals a few days later and wrote "Wow!" next to it on a piece of paper. The signal has, since then, been known as the "Wow! Signal". This signal may have come from extraterrestrial sources. There were several attempts made to explain the signal's origin, both from astrophysical sources and terrestrial and artificial sources, but none have been proven. On the other hand, there is currently no known explanation of the signal's origin that could suggest it was from extraterrestrial communications. For example, the signal has no modulation, which is a communication feature we use on Earth. The direction of the signal is still being monitored to date by SETI, but it has never been detected again.

A mainstream concept that is often linked to alien lifeforms is a UFO: Unidentified Flying Objects. As their name indicates, they are objects seen by people on Earth or in space, that do not resemble or behave like any known object. Talking about UFOs could lead us to think of alien flying saucers. However, the scientific explanation for UFO is already defined in its name —they are simply flying objects that are not identified. Though this does not automatically put them in the extraterrestrial/alien spaceship category. Why would we think of aliens when there are simpler possible explanations? Not to be a killjoy, but 90% of UFO sightings are or can be, explained. Every day, various objects are mistaken for unidentified objects, including wish balloons (balloons released during weddings and birthday parties), aircrafts, drones, high-altitude balloons, satellites, natural and meteorological phenomena, meteors, optical illusions, and many other familiar objects. Many UFO sightings could also be a misconception or an exaggeration by the people who saw them. Drinking a bottle of whiskey can confuse a balloon for a distant flying alien saucer. Alien abduction stories may also fit in the same category. The remaining 10% are real UFOs; they are simply unidentified, and aliens are just one of the several possibilities we have. If I can think of other possibilities, it could include classified military experiments (if the military is trying a new type of aircraft that hasn't been seen before), secret experiments, unknown human-made objects, strange natural phenomena (similar to lightning), strange optical illusions, unknown astrophysical events, other unknown things... and aliens.

Unfortunately, the only thing that is sure for now, is that all our searches for extraterrestrial intelligence or life, have yet to yield confirmed discoveries. The search for extraterrestrial life is one of the most prominent topics in astronomy. Although it has not yielded significant life-changing discoveries yet, it still remains a hot topic in science fiction. In this area, imagination can take us anywhere.

Frank Drake established the Drake equation to estimate the number of extraterrestrial civilizations with developed communication capabilities in our galaxy. The equation considers several factors, like the number of stars and exoplanets, the fraction of habitable planets, the chances of developing life on those planets, the possibilities to develop intelligent life, the fraction of lifeforms that develop communication capabilities through detectable signals, and the lifespan of such civilization. Since all these factors are not precisely known, and their estimation can differ broadly from one point of view to another, the results of the equation can greatly vary depending on the considered estimates. The accepted results of the equation state that there should be between 1,000 to 100,000 civilizations in the Milky Way. The idea of having thousands of civilizations in our galaxy alone is exciting; it is certainly exciting for me. However, one might ask: if there are so many intelligent civilizations in the Milky Way, how come we did not encounter any of them?

The Fermi paradox embodies this idea. It is a conflict between the lack of evidence of extraterrestrial intelligence and the high chance of their existence. There are several hypotheses that explain this. The most straightforward one is that intelligent life is actually rare, or even unique to our planet. Another hypothesis suggests that close civilizations cannot reach us because, like us, they have not yet discovered a way to communicate with us. Other explanations are on the social side, like aliens might not want to communicate for cautious reasons, or they are simply not interested (after all, our way of thinking might not be the norm in the Universe). One of the popular explanations is that a technologically advanced civilization has more chances of destroying itself through civil war, artificial intelligence uprising, or nuclear holocaust, before being able to reach other civilizations. Finally, there are also economic explanations to consider, like the fact that communicating or visiting other species might be very costly and perhaps not worth it. There are many explanations that we can think of.

Other than listening to radio signals, we also transmit signals to outer space. Human civilization has been broadcasting signals into space in an attempt to communicate with aliens. Even if we are not doing it intentionally, some signals that we use to communicate with each other escape the planet. So how hasn't any alien civilization detected those yet and attempted to contact us? It might be just a matter of space. If we only think of our galaxy, it is spread through at least 100,000 ly. We have developed the ability to send radio signals for less than 200 years now. Let's assume 200 years. Like any other electromagnetic emission, radio signals travel at the speed of light. Suppose we started communicating through radio signals 200 years ago, and radio signals travel at the speed of light, this would mean that the oldest radio signal transmitted from Earth is now 200 ly away. This means that this radio signal only has the ability to reach alien civilizations within a radius of 200 ly. This is negligible compared to a 100,000-ly-wide Milky Way. Moreover, if aliens, 200 ly away, just received our signal, it will take another 200 years for their response signal to get back to us. This assumes that they can detect our signs, interpret them, and be interested in communicating with us. A very decent argument as to why we have not been contacted or visited by aliens is that they are simply very far away and unable to reach us. Thinking of space, the distances between star systems are immensely huge, and interstellar travel, as we know it, takes a very, very long time and a lot of energy. Nearby alien civilizations may not be more advanced than us, and if they were, they might not be interested in communicating with us. They might be so advanced, that we are not considered intelligent life to them, and they have no interest in communicating with us. Would you, for example, have an interest in communicating with an ant colony?

To sum up this line of thought, we mention that the Universe is immense, and we have only discovered a small portion of it. Alien life may only be part of the large bits we have yet to discover. After all, our galaxy might contain more than 400 billion planets. We also search the extraterrestrial worlds for signs of life-support elements, like water and oxygen. We only found a few thousand planets until now, and we have only explored some of them for traces of life. Our work is very far from being complete. As Jill Tarter, the former director of SETI, said, claiming that there is no alien life at this stage is like dipping a cup in the ocean and claiming that there are no fish because we did not find any in the cup.

Now, thinking of time, the Universe is 13.8 billion years old. We only became a civilization a few thousand years ago. If we consider that the Universe is only one-year-old, in comparison, civilization would have started only a few seconds ago. The dinosaur extinction would have happened less than 2 days ago. So, what if aliens visited us a few minutes, hours, days, weeks, or months ago?

The Kardashev scale measures the technological advancement of a civilization based on its energy consumption capabilities. We consider today a Type 1 civilization, a civilization that can control the energy of its planet. Type 2 civilization is a civilization that is capable of controlling its star energy. Type 3 civilization is capable of controlling its galaxy energy, and is capable of easy interstellar travel. We can add other types from there, like a Type 4 civilization that is capable of controlling the energy at the scale of the Universe and is capable of intergalactic travel. We are assumed to be a type 0.7 civilization at this stage. Becoming a type 1 civilization might take a few hundred years, type 2 a few thousand years, and type 3 a many more thousand years. It is possible that before we became an advanced civilization, a type 3 civilization tried to communicate with us. They could have also visited Earth, but there was no intelligent life then. An alien civilization, much more ancient and advanced than ours, could have landed on Earth 100 million years ago. At that time, they would have met dinosaurs and other forms of life that existed on the planet. Moreover, as mentioned above, some ideas stipulate that if a civilization is advanced enough to achieve interstellar travel, it would have also developed mass destruction weapons, and annihilated itself by that time.

Intelligent life might have existed, but nobody could reach the other simply because before that, civilization destroyed itself with mass destruction weapons. This is what a lot of people fear for humans. Looking at our history and the wars on Earth, this is not a small idea. I ask some questions to the reader and myself: If time travel is possible, are past, and future visitors, humans that settled on other worlds, considered aliens? An unproven theory says that life might have arrived on Earth, on board an asteroid, that impacted our planet. In this case, wouldn't that make us aliens?

Closing Note

We are the Universe

It seems that it all started with the Big Bang. This was followed by an inflation period, and then an expansion period. In the beginning, it was just a hot soup of elementary particles and radiation. A few hundred thousand years later, the Universe cooled down, and the first atoms of Hydrogen and Helium formed. The Universe, continually expanding, is now mainly clouds of Hydrogen and Helium.

These clouds then came together a few hundred million years later to form the first stars. Stars are the factories of the Universe. In their cores, heavy elements are forged, and what cannot be forged inside the core, is forged when these stars reach their cataclysmic ends.

The first galaxies then formed in the first billion years after the Big Bang, and the process continued. Large structures of the Universe began to unveil filaments and clusters. New galaxies were born, galaxies were merging, and some old galaxies were slowly dying. Some galaxies are spirals, some are elliptical, and others are irregular.

Around 13.6 billion years ago, our spiral Milky Way was formed. Surprisingly, it is one of the oldest galaxies. The large masses of gasses moving here and there also shape our world. New stars are born, and old stars are dying. Dying stars are spreading newly created heavy elements everywhere. Atoms are forming molecules, and new elements are slowly populating the Universe. Transient events here and there are flashing in a universal firework show. Meanwhile, the Universe is expanding.

Five billion years ago, inside our galaxy, in one of the arms, in a region rich in gas, orbiting the galactic center at 27,000 ly from the remnants of dying stars, a new system is formed: The Solar System. The Sun now contains Hydrogen and Helium, and it also contains heavy material previously created by ancestor stars. The planets then formed with the Sun. One of them, Earth, is the 3rd planet.

In the first billion years, the solar system was chaotic, and Earth was too hot and it was heavily bombarded by asteroids. Then things started to settle down, and Earth was cooling down. Orbiting the Sun at 150 million kilometers, Earth has the ideal temperature to sustain liquid water, so water structures started to appear on its surface. Inside these waters, molecules probably came together and created organic compounds which grew to form more complex compounds. More complex compounds then started forming until they could carry information. Eventually, DNA was formed. These are the first forms of cells that can evolve. This might be how life started on Earth. Life evolved to form more complex organisms: from DNA, cells, and multicellular organisms, until the first water animal evolved.

Following the theory of evolution, the animals started evolving over millions and millions of years and somehow evolved into land animals. The ones living by the shore probably started developing land animal features, like legs. Land animals continued to evolve until, eventually, a subgroup evolved into what is known today as humans.

The first human ancestors appeared around 6 million years ago. Homo sapiens, our ancestors, emerged about 300,000 years ago in Africa. Humans continued to evolve and formed the first civilization around 7,000 years ago. Then less than two centuries ago, in the mid-19th century, industrialization began, and technology evolved exponentially. Today, we can travel to space.

We are reading this book on paper, on a computer, on a tablet, on the phone, or we are listening to someone read it to us as an audiobook on some application. We eat, drink, and breathe the same elements from our Earth. Our food comes directly from the Sun. The Sun provides the required energy for the photosynthesis of plants which are the base of the food chain. The molecules in our DNA come directly from this Earth. The molecules in this Earth came directly from the Sun's formation. The molecules and atoms in the Sun, come directly from the molecular gas cloud that has formed our solar system. The atoms in this gas cloud were produced in the heart of the stars that died before the birth of the Sun. The leftovers of older stars formed these new stars. These older stars formed in the hydrogen and helium clouds after the recombination period. The atoms in these gas clouds were assembled by the elementary particles that came after the Big Bang.

All of these were in a singularity, and the whole story started 13.8 billion years ago. Today, we might have different ages, but the truth is that the elements forming our body are 13.8 billion years old. We are made from the same material that everything around us is made of.

When we look at space, we are studying the elements that we are made of. We are looking at our history. We are looking at our home. We are looking at ourselves. We are stardust. We are the Universe looking at itself. We are the Universe taking consciousness. We are the Universe.

Glossary

Objects and phenomena

AGN: Active Galactic Nucleus

BCG: Brightest Cluster Galaxy

CDM: Cold Dark Matter

CMB: Cosmic Microwave Background

FRB: Fast Radio Burst

GRB: Gamma-ray Burst

GW: Gravitational Wave

KN: Kilonova

SN: Supernova

SGR: Soft Gamma-ray Repeater

TDE: Tidal disruption event

UFO: Unidentified Flying Object

VHE: Very High Energy

Missions and rockets

DART: Double Asteroid Redirection Test

ISS: International Space Station

SLS: Space Launch System

Agencies and institutes

CERN: European Council for Nuclear Research

CSA: Canadian Space Agency

ESA: European Space Agency

JAXA: Japan Aerospace Exploration Agency

NASA: National Aeronautics and Space Administration

SETI: Search for Extra-Terrestrial Intelligence

Facilities and infrastructure (observatories)

1HT: One Hectare Telescope

AMS: Alpha Magnetic Spectrometer

ANTARES: Astronomy with a Neutrino Telescope and Abyss environmental RESearch project

ASKAP: Australian Square Kilometre Array Pathfinder

ATA: Allen Telescope Array

ATCA: Australia Telescope Compact Array

ATHENA: Advanced Telescope for High ENergy Astrophysics

CE: Cosmic Explorer

CHIME: Canadian Hydrogen Intensity Mapping Experiment

CTA: Cherenkov Telescope Array

EHT: Event Horizon Telescope

ET: Einstein Telescope

ELT: Extremely Large Telescope

FAST: Five-hundred-meter Aperture Spherical radio Telescope

GAIA: Global Astrometric Interferometer for Astrophysics

GBT: Green Bank Telescope

GLAST: Gamma-ray Large Area Space Telescope

GTC: Gran Telescopio Canarias

HAWC: High Altitude Water Cherenkov

H.E.S.S.: High Energy Stereoscopic System

HK: Hyper-Kamiokande

HST: Hubble Space Telescope

HXMT: Hard X-ray Modulation Telescope

INTEGRAL: INTErnational Gamma-Ray Astrophysics Laboratory

JWST: James Webb Space Telescope

KAGRA: Kamioka Gravitational Wave Detector

KM3NET: Cubic Kilometre Neutrino Telescope

LHAASO: Large High Altitude Air Shower Observatory

LIGO: Laser Interferometer Gravitational-wave Observatory

LISA: Laser Interferometer Space Antenna

LSST: Large Synoptic Survey Telescope

MAGIC: Major Atmospheric Gamma Imaging Cherenkov Telescopes

MeerKAT: Karoo Array Telescope

NuSTAR: Nuclear Spectroscopic Telescope Array

OHP: Observatoire de Haute Provence

SK: Super-Kamiokande

SKA: Square Kilometer Array

SWGO: Southern Wide-field Gamma-ray Observatory

TMT: Thirty Meter Telescope

VERITAS: Very Energetic Radiation Imaging Telescope Array System

VLT: Very Large Telescope

XMM-Newton: X-ray Multi-Mirror Mission Newton

ZTF: Zwicky Transient Facility

Other

AD: Anno Domini

BC: Before Christ

PET: Positron Emission Tomography

PMT: Photomultiplier Tubes

USA: United States of America

USSR: Union of Soviet Socialist Republics

Units

Energy

Joules	J

Power

Watt	W

Voltage

Volt	V

Mass

Kilogram	kg
Ton	t
Solar mass	$M\odot$

Distance

Meter	m
Kilometer	km
Astronomical unit	au
Light year	ly

Parsec	*pc*
Megaparsec	*Mpc*

Frequency

Hertz	*Hz*

Temperature

Degrees (Celsius)	*°C*
Kelvin	*K*

Angle

Arcsecond	*arcsec or "*
Arcminute	*arcmin or '*
Degree	*degree or °*

Credits

Book cover: Image taken by Charbel Zeinati. Image features Emilio El Hajj and Halim Ashkar

Chapter 1 cover: NASA / JPL / Caltech / Malin Space Science Systems/ STScI Hubble Deep Field Team

Chapter 2 cover: NASA / JPL-Caltech / ASU / USGS / Pablo Carlos Budassi / Planck collaboration / ESA / GSFC / Arizona State University

Chapter 3 cover: NASA / JPL / Caltech / ASU / USGS /

Chapter 4 cover: NASA / JPL / Caltech / ESA, the Hubble Heritage Team (STScl / AURA)

Chapter 5 cover: NASA / Ames / JPL / Caltech/ MIT/ UCLA / ESA

Chapter 6 cover: NASA / JPL / Caltech / ESA / M. Durbin / J. Dalcanton / B. F. Williams (University of Washington) / CSIRO/

NANTEN2 / C. Clark (STScI)

Chapter 7 cover: ESA / Hubble / NASA / ESA, the Hubble Heritage (AURA/STScI)-ESA / Hubble Collaboration, and A. Evans (University of Virginia, Charlottesville / NRAO / Stony Brook University) / JPL-Caltech / CTIO / IPAC / Event Horizon Telescope Collaboration / P. Cote (Herzberg Institute of Astrophysics) and E. Baltz (Stanford University) / ESO / DSS / Hubble SM4 ERO Team

Chapter 8 cover: ESA / NASA / JPL-Caltech / Planck collaboration / Hubble / the Hubble Heritage (AURA/STSCI)-ESA / Hubble Collaboration / A. Evans (University of Virginia, Charlottesville / NRAO / Stony Brook University)

Chapter 10 cover: ESA / Pierre Carril / NASA / JPL / Caltech / JSC / Roscosmos

Chapter 11 cover: NASA / JPL / Caltech / ESA / SKA / HAWC / SK / IceCube

Chapter 12 cover: NASA / Ames / JPL / Caltech / NRAO / VLA / Parkinson's Movement

Figure 1.1: NASA / GSFC / Arizona State University